Co-published by IWA Publishing
Alliance House, 12 Caxton Street,
London SW1H 0QS, UK
Tel. +44 (0) 20 7654 5500
Fax +44 (0) 20 7654 5555
publications@iwap.co.uk
www.iwapublishing.com

ISBN: 9781789060812
ISBN: 9781789060829 (eBook)
DOI: 10.2166/9781789060829

Foreword

By Kalanithy Vairavamoorthy
IWA Executive Director

The water sector is experiencing major transitions as global changes affect the ways in which we manage water, particularly in urban areas. Under pressure to efficiently manage scarcer and less reliable water resources, the leaders in this sector from around the world are recognizing the importance of working with nature to solve water-related challenges. These Nature-Based Solutions (NBS) will need to be based on a foundation of research and innovation, coupled with a systems perspective of the watershed.

The International Water Association's (IWA) 5-year strategic plan 2019 – 2024 outlines the Association's commitment to providing a targeted platform that helps utilities share experiences, recognize and learn from emerging disruption, and to adapt and embrace change. In addition, the strategic plan encourages a systems approach to water security, through an understanding of the flows that take place at the boundaries between the different water users or sectors.

Consequently, the Association is well positioned to share best practices from around the world to promote the implementation of sustainable water management. The publication of Nature for Water: A Series of Utility Spotlights highlights innovators and the adopters from across the IWA membership of new approaches to water management as a means of fulfilling this commitment.

One of the greatest challenges to mainstreaming NBS stems from a need to strengthen the knowledge base and improve research and innovation on the topic. The IWA has an opportunity to leverage its member expertise and thought leadership to guide water and wastewater utilities as they consider the benefits of these approaches for water security. IWA is working with utilities and water regulators around the globe to make information about NBS accessible and applicable.

This publication offers evidence that holistic and coherent approaches to solutions can be designed through multi-stakeholder participation. The IWA can facilitate collaboration and provides these valuable partnership building platforms to support the design and implementation of NBS projects.

In partnership with The Nature Conservancy, the IWA will continue to promote knowledge sharing and cooperation to accelerate innovative solutions for water management. IWA hopes that through the Nature for Water publication, it can inspire and support utilities in achieving the transformational shifts that are needed in how water is currently managed.

Foreword

By Andrea Erickson-Quiroz
Managing Director, Water Security

Achieving water security is fundamental to sustainable economic and human development. Yet many water resources continue to be poorly managed. Hundreds of millions of people around the world are still living without clean water or sanitation, while others suffer water-related shocks associated with flooding, scarcity and pollution.

As key actors in the provision of secure and sustainable water to both urban and rural populations, water utilities and operators have a unique role in leading innovative approaches to sustainable and resilient water management. Utilities will need to develop new skills and solutions in order to build flexible systems that can respond to water security challenges posed by changing climates and growing populations.

The good news is there is a powerful ally hiding in plain sight: nature. Managing lands and waters by employing nature-based solutions (NBS) – such as reforestation, protection of riverine riparian zones, and agricultural best-management practices – can contribute to improved water quality and increased flows, while at the same time support local communities and underserved populations. Nature, though often underutilized, offers cost-effective and scalable solutions that can provide important co-benefits like carbon mitigation, improved rural livelihoods and biodiversity gains alongside traditional ecosystem services like filtration and recharge. By establishing nature as a cornerstone of water management, we can transform the water sector for people and nature's benefit.

The International Water Association and The Nature Conservancy are partnering to advance sustainable water management through NBS by supporting those water utilities, operators, and regulators looking to harness nature and natural infrastructure to enhance water security. Nature for Water: A Series of Utility Spotlights contributes to a growing body of knowledge by capturing the lessons learned from utilities representing as they address various challenges and opportunities.

The Nature Conservancy is proud to support and work with utilities and operators who are leading positive change in the water sector. We look forward to continuing the conversation with these change makers and creating a world where people and nature thrive together.

Contents

Reservoir in Alto Cotia System © PEDRO BEICHT

Nature enthusiasts tour a Greenseams® property to see how hydric, or sponge like, soils help reduce the risk of flooding by managing water where it falls. © MMSD

Introduction

There is increasing attention and interest in the use of nature-based solutions (NBS) to manage our water systems, from flood protection to water storage to securing drinking water supply. NBS harness the natural processes that regulate the water cycle to ensure adequate volumes of water of suitable quality can be sustained well into the future. NBS involve approaches to water management that protect, sustainably manage or restore ecosystems to address societal challenges, while simultaneously delivering co-benefits to humanity and biodiversity (Cohen-Shacham, 2016). Implementing NBS can be a cost-effective alternative or compliment to the traditional grey infrastructure approaches used to guarantee water quality and quantity. As one example, source water protection through good land management can improve water filtration, produce more reliable downstream flows, and reduce the amount of sediments and nutrients entering the rivers, springs and aquifers that feed urban water supplies (Abell, R., et al. 2017). As climate change continues to intensify hydrologic cycles, water managers and users will need a diverse set of strategies to cope with the threats posed to their water supplies. A significant advantage of NBS is their ability to achieve system-wide outcomes, meaning they can bring social and economic benefits beyond water-related metrics. NBS present a

valuable opportunity to meet multiple challenges that are central to the 2030 Agenda for Sustainable Development and its 17 Sustainable Development Goals (SDGs).

This idea is further expanded in the 2018 United Nations World Water Development Report (WWDR) which underscores the importance of working with nature to achieve more sustainable management of water resources. The report promotes a holistic approach to water resource management with potential to significantly enhance water security across varied geographical contexts. In light of growing water security challenges from population growth and climate change, the 2018 WWDR emphasizes the importance of working in harmony with nature, rather than against it.

An enabling legal and regulatory environment, financing mechanisms and social acceptance can have an essential role in accelerating the uptake of NBS for water (WWAP 2018). Despite the great potential of NBS to address water-related challenges, contemporary water resource policy and management approaches still have significant room for improvement. Water management frameworks favour dependency on grey infrastructure approaches, placing the onus on utilities to innovate with NBS projects.

Water utilities cite insufficient access to evidence and guidance to decide between grey, green or hybrid options. Regulators and regulatory agencies or authorities1 (including authorities and officials with regulatory and supervisory functions related to the provision of water, sanitation and wastewater management services) don't necessarily recognise or promote NBS as a means of meeting compliance to treat, protect and secure water resources, which heightens perceptions of risk and uncertainty around their performance. To add to these challenges, the lack of awareness and communication between stakeholders, technical guidance and resources, as well as robust performance assessments of existing NBS applications hinder the adoption of this concept in policies.

While local mandates on roles and responsibilities vary, utilities and regulators who play an active role in catchment management are uniquely positioned to support the mainstreaming of NBS. However, unlocking this potential requires adapting the roles and expectations each actor is traditionally expected to fulfil. Utilities prioritize the delivery of water supplies, rather than long-term environmental stewardship. Water resource agencies or basin organizations have mandates to coordinate catchment management initiatives, yet utilities rarely engage in an active role, despite their reliance on the health of the catchment for raw water supplies. The lack of utility involvement in catchment management is a lost opportunity when it comes to mainstreaming NBS, considering their direct access to a large customer base and established mechanism for collecting payments. That said, the idea that water providers can and should be champions of environmental stewardship is gaining increasing acceptance and translating into concrete action. The case studies in this publication suggest that utilities are prioritizing watershed management and catchment protection efforts through partnerships with water resource agencies, basin organizations and their regulators. Many of these utilities report lower operation and maintenance costs and the possibility of postponing or avoiding major infrastructure expansion as a result. These

positive outcomes support the businesses case for utilities seeking to protect the ecosystem services they rely on and must be publicized to encourage action among other utilities. The case studies demonstrate the emergence of a new role for utilities, as they shift from being solely water providers to protectors.

A re-evaluation of the role of regulators is also needed to support the accelerated deployment of NBS. To support an enabling environment for NBS, regulators must evolve from their traditional function as enforcers to one of partners. Regulators are well positioned to identify entry points where consideration of NBS can be incorporated into policies or procurement practices, as well as in the design of financial mechanisms or utility investment plans. Successful implementation of NBS is contingent on interactions and negotiations between multiple, interdependent stakeholders at various governance levels. For example, the SDGs provide an international framework for integrating NBS into relevant national regulations and policies. Recognizing the potential contributions of NBS to the SDGs underscores their significant co-benefits beyond water management and can help in translating these benefits into assessments, cost analyses and policy (WWDR 2018). Wood et al., (2018) found that 44 targets underlying the 17 SDGs focus directly on improving the environment and dimensions of human well-being related to the environment (e.g., health, poverty, nutrition, spirituality). As country governments look to initiatives that support SDG commitments, examining if and how existing regulatory frameworks can be strengthened to incorporate NBS can be an entry point for meeting SDG targets. Integrating NBS within traditional systems and managing risk will require guidance on the part of regulatory agencies. Their central role calls for a partnership-oriented approach to working with water utilities.

The following ten (10) case studies share a utility's perspective on the journey of adopting NBS for water resource management. These cases illustrate the value of enabling factors such as appropriate legal and regulatory frameworks, financing mechanisms and social acceptance can provide for the long-term viability of NBS programs. They also illustrate the potential for a well-designed, community-based NBS concept to spark public engagement and raise awareness for sustained participation. Despite the context-specific nature of NBS applications, many of these experiences demonstrate similar motivations, approaches, conversations, pending needs and frustrations across the utilities interviewed for this work. For example, effective engagement and partnership with the agricultural sector is a common challenge for many utilities striving to protect source watersheds and minimize pollution. From the motivated yet nascent settings where NBS conversation are just beginning to take root, to the advanced, scientifically-backed projects that are informing national policy frameworks, these utilities share a common interest in maximizing nature's potential to enhance water security.

Gathering utility experiences from distinct geographies attempts to present a broad perspective of how water service providers are grappling with water resource challenges under complex and varying circumstances. While utilities in Denmark and the United States are using natural infrastructure to address flooding and stormwater runoff, utilities in Italy and Brazil are undertaking NBS as a strategy for coping with the threat of water scarcity. Many utilities face the pressure of deteriorating water quality as a result of erosion, urban or agricultural pollution. Across the board, water providers are concerned about meeting the demands of their growing populations amidst the intensification of these pressures that will be amplified by the impacts of climate change. In response to water resource challenges, the utilities in this publication have adopted a wide range of NBS interventions. By contextualizing each NBS application within its social, institutional and regulatory environment, the publication endeavours to shed light on the obstacles that hinder scalability and the factors for success.

The current state of global water security calls for a rapid and holistic reassessment of the way water resources are managed. Recognizing the crucial role of utilities and regulators in leading this change, the International Water Association (IWA) and The Nature Conservancy (TNC) have engaged their networks to understand how motivation and concrete action around NBS are taking shape at local levels. The initiative aims to contribute new insights to the body of knowledge against which NBS can be assessed, a critical element for mainstreaming these practices. This compendium of case studies evidences the motivation for and successful application of NBS across geographies and diverse regulatory, financial and cultural environments. These experiences indicate that innovation and leadership at the utility level, particularly when paired with active regulatory involvement, can have a profound impact on the implementation of NBS for water management. Sharing these case studies will assist the development of new programs and support experimentation while expanding the base of interested utilities and regulators.

REFERENCES

Abell, R., et al. (2017). Beyond the Source: The Environmental, Economic and Community Benefits of Source Water Protection. The Nature Conservancy, Arlington, VA, USA.

Cohen-Shacham, E., et al. (2016). "Nature-based solutions to address global societal challenges." IUCN, Gland, Switzerland 97.

IWA (2015). The Lisbon Charter. International Water Association. London, United Kingdom. Available at https://iwa-network.org/wp-content/uploads/2015/04/Lisbon_Regulators_Charter_SCREEN_EN_errata.pdf

Sylvia Wood and Sarah Jones (2017). Landscape news. Laying out a road map for nature-based solutions in the Sustainable Development Goals. *Linking ecosystem services to multiple dimensions of human well-being.* Available at https://news.globallandscapesforum.org/viewpoint/laying-road-map-nature-based-solutions-sdgs/

WWAP (United Nations World Water Assessment Programme). 2018. The United Nations World Water Development Report 2014: Nature-Based Solutions for Water. Paris, UNESCO. Available at http://www.undp.org/content/undp/en/home/librarypage/environment-energy/water_governance/nature-based-solutions-for-water.html

Nature-Based Solutions for Water Management

Major categories of NBS approaches identified in the case studies in this report are represented below as illustrations.

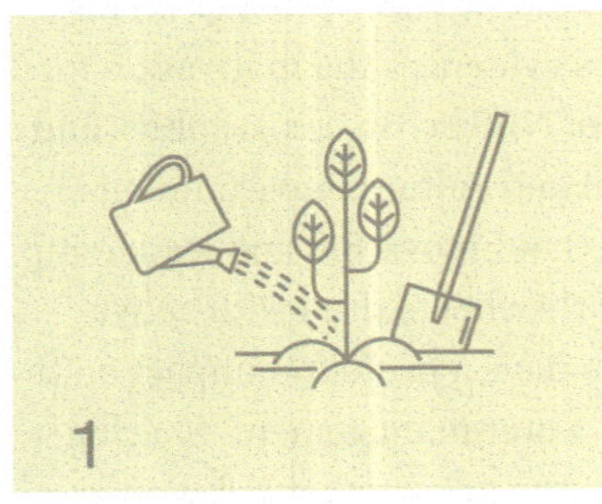

1. Reforestation and forest conservation

2. Riparian buffers or restoration

3. Wetland construction, restoration and conservation

4. Flood bypasses, green infrastructure for flow regulation

5. Urban green infrastructure including green roofs, spaces and water harvesting

6. Targeted land protection, including agricultural best management practices and improved soil health and monitoring

ANGLIAN WATER

LOCATION
West Norfolk, United Kingdom

POPULATION SERVED
6 million

NATURE-BASED SOLUTIONS
3. Wetland construction, restoration and conservation

REGULATORY DRIVERS
European Union Water Framework Directive

Many interested yet hesitant utilities and regulators cite a lack of evidence that nature-based solutions can deliver water quality and quantity results. Chris Gerrard, Natural Catchment and Biodiversity Manager at Anglian Water, argues that there is sufficient information and utility success stories available for that to no longer be an acceptable excuse.

Anglian Water is the largest water and water recycling company by geographic area in England and Wales, and it is located in a region of the country that receives, on average, a third less rainfall than the rest of England (Anglian, 2019). The company is committed to solving environmental problems at source and water quality threats prompted evaluation of nature-based approaches to water management. Contrary to concerns about the cost-effectiveness of NBS, the Anglian Water case study demonstrates how these programs can actually save money for business and customers. Their experience showcases the critical role of customer engagement and how strengthening the relationship between utilities and the general public can create avenues to prioritize natural capital approaches.

How did a company that operates a quarter of all water and water recycling treatment works in England and Wales approach the topic of NBS? The first step in designing an effective program was to gather a scientifically informed body of research on the problems facing catchments in the region. Combining strong scientific evidence with the experience of other utility case studies has allowed Anglian Water to identify effective interventions. This exercise of diagnosing pressures on natural capital stocks, mapping the catchments where intervention was most needed, and then sharing this information with relevant local stakeholders was an essential building block for implementation.

Aerial view of Ingodisthorpe recycling center with ponds forming the new treatment wetland
© ANGLIAN WATER

Strong mandates from the community and local organizations to implement nature-based solutions are a powerful component of the Anglian case study. A clear example of this local pressure is illustrated through the initiative of a local charity focused on catchment-based approaches, the Norfolk Rivers Trust. With financing from Anglian Water, in 2017 the Trust pioneered the construction of a wetland treatment site that improves treated effluent before allowing it to return to the River Ingol. The wetland filters water after it has passed through the existing treatment plant to ensure it meets high quality standards, replacing the need for conventional, energy intensive infrastructure. It is the first of its kind in England. The Trust recognized the potential for this project to generate cost savings, reduce carbon emissions and increase wildlife in the area. It provides an excellent solution to removing unwanted chemicals in a natural way, rather than through additional chemical treatment or infrastructure, which would additionally raise costs for customers. The success of this pilot project gave Anglian Water the confidence to further promote NBS, both internally as a company and externally to industry regulators. The company recently unveiled proposals for the development of dozens of wetlands for wastewater treatment in its next business plan period 2020-2025. In the UK, public pressure for catchment protection extends beyond local environmental organizations. Anglian Water initiated an ambitious and targeted customer engagement strategy to assess the level of support for NBS in their communities. Customers responded positively, encouraging the company to engage in greater awareness raising efforts. "The more we inform our customers about what we do, the more they appreciate the challenges we face as a company and the challenges growth and climate change present to the environment. When they have that level of understanding, we get the highest level of support for NBS, because customers understand the full potential benefits and recognize their role in solving them", Chris comments.

For utilities struggling to prioritize NBS in their corporate agenda, the case of Anglian Water provides a strong evidence base. Regulatory requirements such as the EU Water Framework Directive provided a starting point for Anglian Water to move away from end of pipe solutions and think about water issues at a catchment level. However, it's clear that the growth of Anglian's innovative program goes beyond regulatory compliance. A company commitment to sustainability, community driven pilot studies and customer pressure to deliver environmentally-friendly solutions have been vital elements to support the mainstreaming of NBS.

It is well supported that multi-stakeholder networks are important to the design, planning and implementation of NBS programs. However, water utilities frequently cite tense relations with farmers when catchment protection activities involve resource stewardship or significant financial investment from the agricultural sector. To combat these tensions, Anglian Water employs appropriately qualified advisors to speak to farmers about their practices. Partnership building involves dedicated catchment advisors with an agricultural or farming background forging relationships with farm managers. Initiating conversations in a way that understands the interests, concerns and priorities of these stakeholders has allowed Anglian to achieve its aims for catchment protection. The Slug it Out Campaign (see Box 1) is evidence that this approach has been highly successful. The model of delegating appropriate staff to engage meaningfully with farmers addresses the perception that large water companies are disconnected from the populations they serve. This is a promising strategy for building a strong and engaged multi-stakeholder network.

As Brexit discussions leave a high degree of uncertainty around future water quality standards and compliance in the UK, Chris Gerrard reflects on how the EU's regulatory environment can better support NBS adoption. "We play our part in protecting and improving water quality. Right now we often have to use chemical and energy-intensive treatment to do this; improving the water environment but impacting other elements of the natural world. We'd like to see NBS mainstreamed where we can jointly agree with the regulator it's the best approach for the environment as a whole, even if they don't work as consistently as a traditional solution". While a NBS approach might not deliver the required water quality parameters in a given month or year, there are nature-based metrics indicating co-benefits. The key, and challenge, behind regulation is providing regulators with real data. Anglian Water seeks to devise a new approach with the Environment Agency that allows projects like the wetland wastewater treatment facility to become the rule, rather than the exception. Irrespective of how this materializes, their 2020-2025 Business Plan will build on the success of the past 5 years to further embed natural capital approaches into decision-making.

Aerial view of Ingodisthorpe recycling center with ponds forming the new treatment wetland © ANGLIAN WATER

For utilities struggling to prioritize NBS in their corporate agenda, the case of Anglian Water provides a strong evidence base. Regulatory requirements such as the EU Water Framework Directive provided a starting point for Anglian Water to move away from end of pipe solutions and think about water issues at a catchment level.

However, it's clear that the growth of Anglian's innovative program goes beyond regulatory compliance. A company commitment to sustainability, community driven pilot studies and customer pressure to deliver environmentally-friendly solutions have been vital elements to support the mainstreaming of NBS.

REFERENCES

Anglian Water Group Ltd. 2019. Available at https://www.anglianwater.co.uk/environment/our-commitment/our-plans/slug-it-out.aspx

Anglian Water Group Ltd. 2019. Available at https://www.anglianwater.co.uk/news/slug-it-out-year1.aspx

DE WATERGROEP

"We want as much nature as possible in our catchments. That's our goal. We see that the places where we have only natural activities in the groundwater catchment areas have good water quality, even if it's a very vulnerable aquifer. In extraction areas where we don't have nature, the groundwater quality is not as good and extra treatment is necessary."

Simon Six, Water Resources and Basin Management Team Leader, De Watergroep

LOCATION
Flanders, Belgium

POPULATION SERVED
3.2 million

NATURE-BASED SOLUTIONS
2. Riparian buffers or restoration

6. Targeted land protection, including agricultural best management practices and improved soil health and monitoring

REGULATORY DRIVERS
European Union Water Framework Directive

Flemish Environmental Permit Statute

De Watergroep is the largest drinking water supplier in the Flanders region of Belgium. To serve a customer base of approximately 3 million inhabitants, the utility draws water supplies from 85 groundwater pumping stations and 5 surface water pumping stations. In the densely populated and cultivated region of Flanders, investing in the long-term protection of these water supplies through nature-based solutions (NBS) is a means of addressing serious water quality issues stemming from agricultural and industrial pollution. De Watergroep tackles pollution threats by focusing on the protection and enhancement of the ecosystems that surround their abstraction areas, an approach they have used since 1985. The water utility currently faces restrictions on their surface water pumping activities during the warmer months of March to September. This is attributed to the increased threat of nutrient and pesticide infiltration into surface water supplies and a diminished dilution of chloride coming from industrial discharges. During seasons with low water levels or times of prolonged drought, any subsequent rainfall flushes large amounts of nutrient runoff into the surface water. As the utility with the largest amount of shallow water extraction sites and surface water intakes in the region, the need to protect water resources carries strategic importance for De Watergroep.

Aerial view on the water basins at Kluizen near Ghent for the collection of surface water for drinking water supply. © DE WATERGROEP

Catchment protection in Flanders is recognized as the responsibility of the government, rather than the water company. Utilities have limited impact on land use in the catchment and therefore, are afforded limited responsibility. While these decisions fall under the jurisdiction of local and regional authorities, water utilities are increasingly relied on for consultation and guidance. De Watergroep has secured this role by actively engaging with government agencies, fellow water utilities, research organizations and the local farming community on the topic of water protection. They are invested in developing sustainable solutions that encourage and rely on the involvement of all stakeholders within their catchment areas. To many water utilities facing challenges in combating the impact of agricultural pollution on raw water supplies, De Watergroep's case illustrates how a strong utility voice has significant potential to foster acceptance of a new set of solutions.

Outdated and weakly enforced water protection laws signalled a need for Flemish water utilities to be proactive in addressing the risks to drinking water supplies from inadequate protection. In 2015, in close collaboration with the Flemish Environmental Agency (VMM) and a consortium of Flemish public water utilities, De Watergroep led the development of a risk-based protection plan to control the risks to drinking water supplies and ensure groundwater quality and quantity (Six et al. 2015). An expert group prepared a list of the possible hazards for groundwater quality and quantity. The water

companies used this list to develop a matrix for classifying threats and define appropriate management plans. The strategy was officially recognized in Flemish Legislation in 2013 (Belgian Act, 2013). De Watergroep additionally participates as a member of the Coordination Commission on Integrated Water Policy (CIW), a committee designed to guarantee a coordinated approach to water policy in Flanders. The CIW plays an important role in the planning and implementation of water policy at the river basin level, ensuring alignment with the goals of the Water Framework Directive. De Watergroep utilizes this platform to engage with government agencies and advocate for stronger catchment protection efforts. De Watergroep's proactive engagement merits attention in the NBS space not only because it enables more effective definition of the site specific and appropriate NBS intervention, but because it sets the stage for regional water utilities to build stronger relations with the governmental organizations and other stakeholders involved in water management.

In Belgium, certain drinking water abstraction areas are subject to protection as a means of ensuring quality and reducing treatment costs. Water companies have control over any activities that take place in the most immediate zone surrounding the catchment (defined as zones where water can reach the catchment area within 24 hours). Regulatory requirements like this and the European Water Framework Directive have served as an icebreaker to initiate greater interaction between utilities and regulators. De Watergroep was successful in capitalizing on this opportunity to bring a utility's voice into conversations with regulators and government agencies regarding the governance mechanisms that can ensure protection of water resources. They promote deeper engagement and collaboration with the Flemish government on all matters related to the protection of drinking water, from the design of legislation to communication, to the steering and monitoring of research studies. The Flemish Ministry for the Environment (VMM) now consults them on the design of more effective water quality standards and granting of environmental permits that influence water quality and quantity.

Simon Six, Water Resources and Basin Management Team Leader at De Watergroep, is optimistic given the noticeable improvements they've seen from using a NBS approach to catchment management. De Watergroep has observed the benefits of NBS on water quality for some time, and now sees greater acceptance among stakeholders that NBS can be used as a tool for water protection. At pumping sites, utility staff measure groundwater depth and quality to model the basin and identify areas that would be suitable for NBS interventions. Resulting data from these interventions indicates that wells protected by natural areas have much better water quality compared to areas of intense land use, even in the region's most vulnerable aquifers. In fact, the wells with natural surroundings had little to no parameters indicating diffuse pollution of the aquifers. Simon explains that protecting the quality of the catchment area around wells is a practice that has been in place since pumping activities began in the late 19th century. When the Flemish legislation on groundwater protection was put in place in 1985, the focus on protecting groundwater catchments intensified. Despite legislative backing, challenges remain in conveying these benefits to the agricultural industry and convincing them to adopt best management practices that reduce pollution.

Historically, there has been little interaction between water suppliers and the farming community. Local farmers place value on every square meter of land. This would be a sizable request. However, recognizing the pressures faced by the agriculture sector was important to understanding why farmers have traditionally only been concerned with meeting compliance. Several meetings with farmers have indicated it will be difficult to convince them to take action beyond legal requirements. To make matters more complicated, there are no legal protection zones for surface water. Nevertheless, De Watergroep's efforts to find a common ground with the agriculture sector are generating positive reverberations. The utility is currently designing an informal cooperative program in partnership with the farming community to build the foundation for better dialogue. By modelling groundwater flow to determine which fields have the greatest impact, De Watergroep will identify willing local participants for a pilot project that evaluates how farmers can evolve toward more sustainable agricultural practices. The pilot project will be fundamental to developing a multi-stakeholder dialogue that discusses new solutions for agricultural pollution.

Moving forward, De Watergroep is actively searching for new opportunities to finance their NBS programs and build a stronger evidence base for their implementation. Pilot projects are currently financed by the utility, project partners, research grants and a portion of consumer bills dedicated to water protection projects that focus on herbicides and pesticides. De Watergroep also has a small capacity to work with nature management organizations and make strategic land purchases surrounding their wells. Their biggest opportunity, which has yet to be brought into practice, lies in the possibility of directing yearly groundwater taxes paid to the government toward water protection efforts.

De Watergroep's NBS approach to groundwater and surface water protection was primarily driven by factors that are common to most water utilities around the world: concerns over water quality and regulatory compliance. The utility is dedicated to mapping opportunities for strategic partnerships and collaborations with those invested in the health of Flanders' catchments. "The best way to convince the people is to have good pilot cases, a strong and clear presentation of the problem and opportunities for the different sectors to meet each other. We're constantly looking for those opportunities."

Heads of shallow wells in a managed alluvial forest © DE WATERGROEP

REFERENCES

Belgian Act 2013, Energy and Environmental systems for the Flemish region (EMIS) (2015). Act of the Flemish Government of amending the act of the Flemish Government of 13 December 2002 on the quality and supply of water for human consumption, as to impose public service obligations on risk management, crisis management and supply security). Publication of 9 December 2013, pp. 97477–97480. https://emis.vito.be/sites/emis.vito.be/files/legislation/migrated/sb280103-4.pdf (Accessed June 10th, 2019)

Coordination Commission on Integrated Water Policy. Integrated Water Policy in Flanders. (2003)
Available at: http://www.integraalwaterbeleid.be/en (Accessed June 26, 2019)

Six, S., Diez, T., Van Limbergen, B., & Keustermans, L. (2015). Development of a risk-based approach for better protection of drinking water catchments in Flanders (Belgium). *Water Science and Technology: Water Supply, 15(4)*, 746-752. https://iwaponline.com/ws/article/15/4/746/27545/Development-of-a-risk-based-approach-for-better?searchresult=1

COMPANHIA DE SANEAMENTO BÁSICO DO ESTADO DE SÃO PAULO (SABESP)

"Society still does not see watershed conservation among the responsibilities of a water and sanitation company... The utility's role is still largely perceived as implementing sewage and water infrastructure yet focusing on the long-term protection of water supplies ensures the sustainability of our business."

Mara Ramos, Manager of Metropolitan Water Resources, SABESP

LOCATION
São Paulo, Brazil

POPULATION SERVED
26.7 million

NATURE-BASED SOLUTIONS
1. Reforestation and forest conservation

2. Riparian buffers or restoration

REGULATORY DRIVERS
State level regulatory mandates

Situated in the population dense, water scarce and wildfire prone area of São Paulo, Saneamento Básico do Estado de São Paulo (SABESP) faces increasing pressure to supply sufficient and good quality water to its 26.7 million customers. In 2015, Southeast Brazil grappled with its worst drought in nearly a century. Increased rainfall in the following years provided some relief to water and sanitation suppliers like SABESP, but threats to long-term water supplies will likely continue. To protect and restore their catchment areas from degradation, SABESP designed an NBS program that prioritizes reforestation and engages water users in coordinated conservation efforts across the watersheds it depends on.

In the São Paulo Metropolitan Area, the Cantareira system is the most important watershed. Composed of six reservoirs, Cantareira is one of the largest water supply systems in the world and responsible for supplying water to almost half the city of São Paulo. The wellbeing of this area is highly jeopardized by land conversion to support agricultural, pastoral, industrial and urban expansion. These activities have already consumed over 70% of the original forests that comprised Cantareira's watershed areas (Abell et.al, 2017)

SABESP's catchment protection efforts stem back to the 1980s and were focused on protection and restoration of SABESP's own property areas. The 2015 drought, combined with the potential for prolonged strains on water supplies, prompted the development of a new program in 2017 called Cinturão Verde dos Mananciais Metropolitanos (Green Belt of Metropolitan Watersheds). To design a new approach to water management, SABESP's Water Resources Department drew on experiences from The Nature Conservancy's São Paulo Water Fund, technical studies on the value of investing in natural infrastructure in São Paulo (Ozment et al, 2018), and the work of local academic institutions. The Cinturão Verde program delivers on SABESP's mandate to go beyond the traditional utility role of protecting owned assets to adopting collective action approaches across the catchment for ensuring water security.

SABESP's catchment protection program utilizes nature-based solutions like reforestation and revegetation to protect surrounding watersheds, principally those of the vulnerable Cantareira system. The first component of the program requires the preservation of 33,000 hectares of land around four watersheds, Cantareira, Rio Claro, Alto Cotia and Capivari. These watersheds equate to 1.4% of all Atlantic forest in São Paulo state. The second mitigates the threat of encroaching urbanization on watershed lands by promoting increased vegetation cover around the Cantareira reservoir. Establishing or conserving these forests can protect water supplies by regulating sediment flow and filtering or preventing the entry of pollutants into waterways. The final component supports the operation of plant nurseries, where native Atlantic Forest and Cerrado species can be cultivated for restoration projects (SABESP, 2018). This year, SABESP will develop a new operational model that involves the participation of a local NGO.

The utility experienced some setbacks in the implementation of their Cinturão Verde program due to wildfire threats. Maintenance of conserved and restored areas has proven difficult as prolonged drought in the region created favourable conditions for wildfires to flourish. SABESP lost 20% of a replanted and protected forest area to wildfires. This expensive setback required them to reconsider their calculations for cover loss and anticipate higher margins for the future restoration initiatives.

SABESP's experience exposes how a focus on short-term water use amidst these pressures has prevailed over development of a long-term strategy. "Water operators don't necessarily consider buying areas of land to protect it or keep it from future urbanization. They look for the right to use the water", comments Mara Ramos, Manager of Metropolitan Water Resources at SABESP. There is a need for a paradigm shift in the way utilities think about their role in watershed protection. Along with this comes a need for cooperative and integrated efforts that join forces from diverse water user groups.

SABESP sees the challenge in prioritizing NBS in regulatory and civil society agendas further deepened by the prevailing view in the political arena that water utilities are simply responsible for raw water extraction and supply, not watershed protection. "Society still does not see watershed conservation among the responsibilities of a water and sanitation company. It's difficult to put this kind of project on the agenda. The utility's role is still largely perceived as implementing sewage and water infrastructure yet focusing on the long-term protection of water supplies ensures the sustainability of our business."

Awareness and action continue to grow at a local level. Since 2005, local institutions in the Mantiqueira Atlantic Forest (located to the southeast of São Paulo) have worked in multi-stakeholder partnerships to protect the watersheds that supply water to the Metropolitan region (Benini, 2019). Watershed Committees in São Paulo state are allocating public funding sources toward comprehensive watershed conservation. New models are emerging that help enable utilities like SABESP to be a partner in local efforts alongside regulatory actors. São Paulo's regulatory sanitation and energy agency, ARSESP, has included the topic of NBS in their regulatory agenda and is collaborating with TNC to develop a watershed conservation workplan. With plans to review their water tariff structure in 2021, ARESP has expressed interest in learning how and if NBS can be integrated into their tariff formula. TNC is facilitating the exchange of knowledge

Cattle grazing near Nazare Paulista, Brazil. © SCOTT WARREN, Courtesy of The Nature Conservancy

and experiences by connecting ARSESP with regulatory agencies like Agência Reguladora de águas, Energia e Saneamento do Distrito Federal (ADASA), Agência de Regulação de Serviços Públicos de Santa Catarina (ARESC) and Empresa Municipal de Água e Saneamento (EMASA) as they venture into this multi-year evaluation process. Given their investments and enthusiasm in NBS, SABESP will be a key collaborative partner and stakeholder to consult.

Finding new ways to quantify the economic and ecological benefits of SABESP's program and disseminate these findings on a large scale are the next steps to creating greater awareness of the dependence of cities and economies on their supplying watersheds. SABESP recognizes that mobilizing greater investment for NBS will be dependent on an attractive business case. Access to resources and frameworks that can help evaluate the business case for specific contexts is necessary to move the conversation forward. In the meantime, Mara Ramos and the Water Resources Department will continue to find spaces where NBS programs are shared and evaluated, as well as support the efforts of other water and sanitation companies using NBS to solve their water resource challenges.

REFERENCES

Abell, R., et al. (2017). Beyond the Source: The Environmental, Economic and Community Benefits of Source Water Protection. The Nature Conservancy, Arlington, VA, USA.

Benini, Rubens. (2019). The Nature Conservancy Brazil & Latin America. Roots for Growth: Why protecting and restoring forests is one of the best thing any government can do for its people. Available at https://www.nature.org/en-us/what-we-do/our-insights/perspectives/protecting-restoring-mantiqueira-forest-brazil/.

Ozment, S., Feltran-Barbieri, R., Hamel, P., Gray, E., Ribeiro, J. B., Barrêto, S. R., & Valente, T. P. (2018). Natural infrastructure in São Paulo's water system. Natural infrastructure in São Paulo's water system.

SABESP (2018). Beyond water. *The environmental preservation initiatives in one of the world's largest urban areas, the Metropolitan Region of São Paulo.* Second edition. November 2018. Available at http://site.sabesp.com.br/site/muitoalemdaagua/ (Accessed June 26, 2019)

QUEENSLAND URBAN UTILITIES (QUU)

"We had to change the way we operated with regulators. We had to change our perspective of the regulator as an enforcer to that of a partner, and an integral part of our business."

Paul Belz, Executive Director, QUU

LOCATION
Queensland, Australia

POPULATION SERVED
1.4 million

NATURE-BASED SOLUTIONS
2. Riparian buffers or restoration

4. Flood bypasses, green infrastructure for flow regulation

REGULATORY DRIVERS
Environmental Protection Act 1994 (Queensland)

Queensland Urban Utilities (QUU) is a provider of drinking water, recycled water and sewerage services to a population of over 1.4 million customers in South East Queensland (QUU, 2018). QUU's commitment to serve as a leader for environmental innovation is evidenced by their long-standing experience with nature-based solutions. Their leadership reinforces the critical and valuable role that water utilities can play in mainstreaming NBS, as well as the growing evidence base that these approaches can be cost effective options for improving water quality. Most importantly, the QUU case study is an example of how NBS pilot projects rooted in strong scientific evidence can make a promising case for policy reform.

Faced with pressure to effectively service a growing population and address significant sediment and nutrient pollution, QUU capitalized on an opportunity to explore a new approach to catchment management. The utility's Beaudesert Sewage Treatment Plant was in need of an upgrade to mitigate elevated levels of nitrogen load discharged to the Logan River due to decreased recycled water demand during wetter than average years. Natural channel erosion taking place on the Logan River during these wetter years was also contributing to significant sediment and nutrient pollution into the Logan River. The standard utility approach in these circumstances would involve advanced capital works and investment in expensive sewerage treatment plant upgrades, in this case an AUD $8 million upgrade to construct a sequential batch reactor (SBR) for nitrogen removal (WSAA, 2017). These upgrades, along with other hard engineering options to reduce pollution, have very long asset lifetimes, little flexibility and the costs are borne by utility customers. In Australia, most sewage treatment plants are individually regulated by State Government environmental protection agencies. With population projections, the risk of keeping plants and individual operating licenses compliant increases over time. QUU cites investments of around $AUD 500 million to upgrade their fleet of sewage treatment plants to biological nutrient removal processes over the past 20 years. Regulatory drivers or incentives for pursuing alternative investment options like NBS were non-existent until recently.

In 2013, QUU initiated conversations with their regulator, Queensland Department of Environment and Heritage Protection (DEHP) around a voluntary market-based mechanism for nutrient management in degraded river catchments. Rather than investing in costly treatment plants, they proposed that a green infrastructure solution could significantly reduce streambank erosion, thereby reducing sediment and nutrient loads to the river. QUU would utilise the nutrient pollution credits generated by the project to offset nutrient emissions from the sewage treatment plant and ensure licence compliance (WSAA, 2017).

A crucial exercise to making the business case for NBS involved quantifying the annual loads of sediment, nitrogen and phosphorus that were mobilized into the waterway during wet weather events. Streambank restoration projects generally focus on less impactful metrics for water quality, like how many trees are planted or length of streambank restored, and fail to recognize the real downstream environmental impacts from sediment and nutrient pollution. The modelled nutrient and sediment loads (Bank Stability and Toe Erosion Model [BSTEM]) mobilized were so high that people could finally recognise the severity of the issue and understand why the rivers were appearing so dirty.

QUU's proposed approach involved restoring around 500 metres of severely eroded streambank by changing the angle of the riverbank at the escarpment, installing a wooden pilefield, placing rock barriers at the toe of the bank and planting about 7,000 trees, shrubs and hedges to improve riverbank's stability (IWA/TNC, 2018).

Streambank degradation from natural erosion at QUU's Beaudesert pilot project site © QUU

Feasibility studies indicated it was actually more cost effective to adopt an NBS approach over the conventional option of building or upgrading the sewage treatment plant. Less tangible benefits include increased biodiversity outcomes such as a new wildlife corridor, lower chemical usage and GHG emissions and environmentally resilient agricultural and alluvial plains. The resulting voluntary nutrient offset pilot project prevented more than 11,000 tonnes of sediments from entering the river each year due to natural erosion. This equates to an aversion of 5,000kg of nitrogen and 8,000kg of phosphorus loads into the river each year. QUU was able to save $AUD 7 million in capital costs; and $AUD 1 million per year in operational costs compared to the recommended sewage treatment plant upgrade option, while maintaining compliance with nutrient discharge limits at the Beaudesert Treatment Plant.

With limited knowledge and experience with regulated point source related NBS projects, QUU relied on their positive relationship with environmental regulatory agencies to convince them to adopt a risk sharing approach to pilot this beta project. Initial discussions with the regulator focused on the macro issues, such as the relatively small nitrogen load contribution the sewage treatment plant made to the Logan River nitrogen budget, the high level of investment required to comply with licence conditions and the resulting high costs to customers. The idea was to design and deliver a project that could be used as a living laboratory, then perform quarterly monitoring on performance to evaluate the long-term sustainability of NBS.

The pilot project performed extremely well by withstanding a number of flood events over the initial five years of NBS operation, giving regulators the confidence to create a State policy for point source nutrient offsetting. Transitioning from a voluntary mechanism to a State policy (due for public release before July 2019) will allow for more aggressive implementation of NBS in South-east Queensland and the Great Barrier Reef in Central and North Queensland. This engagement with regulatory agencies on the topic of NBS seems relatively smooth and effective compared to the typical experience of water utilities around the world. QUU admits that a change to a proactive mind-set, leadership and strong personal ties with their regulator helped, yet finalizing an accurate and reputable scientific case was challenging. Environmental regulators prefer cautionary approaches and want to see a strong scientific foundation. In early conversations, QUU admits they adopted somewhat of a "transformational" attitude, which has persisted due to its effectiveness but is now evolving into early stage partnerships on certain issues. "We have a lot of upfront discussions on these issues and work through the challenges before we get into detail on concepts, so that we have everyone in the room and on the same page" explains Paul Belz, Executive Director at QUU. It's necessary to strike this relationship and ensure that the regulator is aligned on outcomes. If regulators are not keen to innovate or adopt smart regulations, it will always be a struggle. Fortunately, the Queensland Department of Environment and Heritage Protection (DEHP) were receptive, creating an environment for QUU to test and evolve their pilot project. There are still some barriers on the science side, specifically in terms of understanding nutrient equivalency and how to link diffuse source nitrogen abatement projects to point source nitrogen emissions. QUU is spearheading R&D projects in this space to build scientific evidence and increase certainty around these trading ratios, with the aim of making future approvals easier.

Securing regulatory approval and participation involved changing the way the QUU operated with regulators. "We had to change our perspective of the regulator as an enforcer to that of a partner, and an integral part of our business" says Paul Belz. The relationship between utilities and regulators is often adversarial, accompanied by a lack of trust and no risk sharing. A crucial factor in building the utility-regulator relationship needed to support NBS was inviting members of regulatory departments and state government stakeholders to participate in on-site workshops, so they could witness first-hand the deterioration of streambanks, discuss erosion issues with the land owners and see upstream catchment areas. "Every time we speak about the project, the key part is including the state government department as one of the project partners. When the project receives award nominations, the regulator has been there as part of the team. I think they rightfully and truly feel part of the outcome" adds Cameron Jackson, Leader Water Quality and Environmental Planning at QUU.

Stakeholder engagement in the pilot project has been beneficial to spreading awareness about the importance and benefits of alternative approaches like NBS to protect water quality, land and enhance biodiversity. Landowners along the Logan River had a direct stake in project outcomes; they were losing significant tracts of property with banks eroding at a rate of up to 1 meter per year in some locations. Generating local interest and support for measures to reduce erosion was easy, even in light of the fact that landowners would have to cede 10-15 meters of riparian land to the project. QUU also partnered with local natural resource management organizations to deliver the project and undertake landholder and community engagement. These organizations offered strong relationships with landowners, valuable insights on siting for pilot projects, and were able to secure project approvals for working in waterways and flood zones in a timelier manner than the utility itself.

This successful case study on NBS can, at its root, be attributed to an attractive business case. The utility was able to illustrate upfront savings of $7 million in capital costs and lower annual operational costs, resulting in savings of $5 million over the lifespan of the 10-year nutrient offset period. The offset project also provided a flexible 10 year planning window to better understand actual population growth in the regional city of Beaudesert, complete more thorough sewerage master

New wildlife corridor © QUU

planning, and improve scientific knowledge on NBS. While the financials were impressive, it's clear that the success of this project cannot be attributed solely to financial numbers. QUU's commitment to aligning operational efficiency with environmental sustainability and collective action approach to solving water quality problems is a powerful reminder that a utility's role is much more than just a provider. The Beaudesert Sewage Treatment Plant pilot project has chartered a path for numerous new projects in Queensland focused on partnering with innovative regulators and local natural resource managers to deliver offsite nutrient reduction projects that achieve better water quality and biodiversity outcomes, while delivering cost savings to water utilities and their customers.

REFERENCES

QUU (2018). Queensland Urban Utilities. *Who we are* https://www.urbanutilities.com.au/about-us/who-we-are (Accessed May 23rd, 2019)

WSAA (2017). Water Services Association of Australia. Case study 6 Using nutrients offset to improve the Logan River. *Queensland Urban Utilities* https://www.wsaa.asn.au/publication/next-gen-urban-water-case-study-6-using-nutrient-offsets-improve-logan-river (Accessed May 21st, 2019)

EMPRESA PÚBLICA METROPOLITANA DE AGUA POTABLE Y SANEAMIENTO (EPMAPS)

"NBS is the only way to preserve water for the future generations. This initiative can be replicated across the country if the water regulator shares the same vision."

Marco Antonio Cevallos, Former General Manager, EPMAPS

LOCATION
Quito, Ecuador

POPULATION SERVED
2.6 million

NATURE-BASED SOLUTIONS
1. Reforestation and forest conservation

6. Targeted land protection, including agricultural best management practices and improved soil health and monitoring

REGULATORY DRIVERS
Metropolitan Ordinance 213

Quito has the highest drinking water coverage (99.3%) in Ecuador. The city's water supply is derived from the Andean mountain systems, including protected grasslands known as páramos. 70% of this water supply originates in three protected areas and their surrounding zones; the Cayambe Coca National Park, Cotopaxi National Park and the Antisana Ecological Reserve (IUCN and IWA, 2014). Despite the protected statuses of these Reserves and National Parks, unregulated development and poor agricultural practices are degrading Quito's watersheds. The extraction of natural resources, deforestation and the burning and grazing of páramos reduce soil moisture content and exacerbate erosion. Historically, there has been a lack of resources dedicated to the operation and protection of these areas, which threatens the stability of the ecosystem services they provide (Echavarria, 2011).

As one of the key economic zones in Ecuador, reductions in the yield and quality of water resources delivered to Ecuador's capital city have implications for economic development. This concerned the municipal utility, Empresa Pública Metropolitana de Agua Potable y Saneamiento (EPMAPS; formerly EMAAP-Q), who holds responsibility for providing drinking water and sewerage services to the Metropolitan District of Quito. EPMAPS saw great value in incorporating nature-based solutions into catchment management approaches to improve water security for a growing metropolitan population. They played a leading role in the establishment of a Water Fund, a collective financial mechanism that strengthens integrated watershed management and governance through the financing of conservation actions. The fund is known as Fondo Ambiental para la Protección de las Cuencas y Agua (FONAG). This kind of multi-stakeholder framework for source water protection is especially crucial for accelerating NBS where regulatory environments are nascent, such as the case of Ecuador.

In 1997, The Nature Conservancy (TNC) and partners began negotiations with the Municipality of Quito and EPMAPS to evaluate sources of sustainable financing for natural infrastructure investments surrounding the city. The idea was to create a consumption fee for those who benefit from the water resources in these areas. The consumption fee would fund conservation projects and watershed management for Quito's upstream watersheds, with the goal of sustaining and improving their functions as natural regulators of water quantity and quality. An initial analysis was prompted by several national and international organizations, who recommended dedicating a portion of water bills to the city of Quito for management of source water areas.

FONAG was initiated with support by the U.S. Agency for International Development and endorsed by the Quito Metropolitan Mayor´s Office through EPMAPS (Water Funds, 2017). It is designed as an endowment fund that receives money from public utilities, private companies and non-government organizations. It is "non-depleting" in that the original endowment is not invested in green infrastructure. Rather, the financial returns made upon this endowment are used to finance nature-based solutions for water conservation. FONAG was established as an 80-year contract between TNC and EPMAPS with an initial seed fund of $USD 21,000 (Herndon, 2014). The Inter-American Development Bank provided important early support, followed by several downstream users who later joined as capital contributors, including the public electricity company Empresa Eléctrica Quito (EEQ), Cervecería Nacional (today part of ABInBev), Tesalia Springs (today part of CBC) and Consorcio CAMAREN. The fund is managed by an independent financial institution and is under compliance with Ecuador's private sector regulator. It is overseen by a Board of Directors, which is chaired by the General Manager of EPMAPS. Although the fund is independent from the government, it cooperates with environmental authorities to ensure alignment with government programmes and policies.

FONAG finances the development of a watershed management plan and carries out projects and programs to achieve this goal. Activities are centred around the goals of strengthening alliances, involving various stakeholders, promoting environmental awareness and developing systems of governance. EPMAPS has worked closely with FONAG to protect 20,000 hectares of páramos. FONAG engages in conservation and vegetation recovery activities to protect the catchment with local communities carrying out the planting. Native tree and shrub species with beneficial root systems are used to bind the soil and support water infiltration. Investments have also been made in surveillance programs and water and soil monitoring stations to track the water balance in Quito's upstream watersheds. This data is used to analyse the impacts of green infrastructure initiatives and better target areas for growth.

The impacts of these activities are measured against the fulfilment of the indicators within FONAG's strategic plan and presented every three months to FONAG's Board. The main indicator is progress towards the preservation of 155,100 hectares by 2080 in priority source water areas outside of the National Park system. Current estimates indicate the fund is at 15.75% (24,425 hectares) completion. Other indicators include hectares of restored land, coverage of hydrometeorological data network, number of trainings, and scope of environmental education.

Once the fund was established, there was a need to guarantee a continued source of investment. In December 2006, Metropolitan Ordinance 213 (originally Metropolitan Ordinance 199) was passed, requiring EPMAPS to provide a permanent contribution of 2% of their annual budget to the fund. By 2013, financial contributions increased to approximately USD $1 million per year (IUCN and IWA, 2014). FONAG now has an endowment of nearly USD $18.7 million and an annual budget of USD $2.5 million. EPMAPS is FONAG's largest financial contributor, providing nearly 88% of the Fund's capital.

A local family's organic garden located in the highlands of the Quito watershed
© ERIKA NORTEMANN, courtesy of The Nature Conservancy

Aside from Municipal Ordinance 213, regulatory frameworks to incentivize NBS for water utilities are largely absent in Ecuador. This is a product of Ecuador's historically fragmented water governance framework, which only recently underwent reforms to transition into a coordinated body of actors (Bréthaut and Schweizer, 2018). At the time of FONAG's establishment, an Ecuadorian regulatory body specifically dedicated to water did not exist. The National Water Secretariat (SENAGUA) was only established by Executive Decree in 2008 and the Agency for Regulation and Control of Water (ARCA) was created in 2015. SENAGUA is responsible for sustainable and integrated water management and promotes policies for watershed protection, with an emphasis on the conservation of native forests and paramos, and maintenance of water quality at the source rather than through water treatment. ARCA is concerned with water pricing across the country.

At this stage, there is minimal collaboration between the national regulator and the utility with respect to FONAG's activities. Municipal Ordinance 213 demonstrates engagement at the local level to ensure compliance with the agreed parameters for water service fees. However, activities in water source areas located outside of EPMAPS' jurisdiction require authorization from SENAGUA, as well as meaningful engagement with the people residing in these watersheds. There is a demand for collaborative platforms that can activate partnerships

beyond the boundaries of the utility's jurisdiction. Ecuador's recent legal reforms have the potential to promote improved water governance and resource management in the country. However, there is still much work to be done to solidify collaborative partnerships between local governments, community organizations and the providers of drinking water and sanitation services. FONAG is making important strides forward by leading processes that promote dialogue and consensus among these actors as well as appropriate decision-making methods.

FONAG has celebrated many successes since its establishment in 2000. However, creating a new culture around water management among water users was not an easy process. It was particularly difficult to encourage behaviour shifts among local communities. Agreements that provide financial or technological compensation and trainings to build expertise around other income generating activities have been created for local participation and engagement. Topics for these trainings include ecological tourism, sustainable agricultural and livestock production, improvement of local access to clean water, and improvement of wastewater treatment and source water protection. The fixed income under FONAG guarantees continuity of these activities and assures communities they will receive continued support and collaboration.

Water funds are now sprouting up throughout Latin America in countries such as Colombia, Mexico, Brazil, Guatemala, Peru and Colombia. The growth of Water Funds throughout Latin America has given EPMAPS opportunities to engage with other utilities across the region to share their knowledge and best practices through technical assistance, capacity building, and investment.

EPMAPS acknowledges the need for capacity, resources and delegated responsibility to ensure the viability of a Water Fund like FONAG. In the case of Quito, the municipal ordinance guaranteeing funds from EPMAPS to FONAG has been essential. As the Fund develops its activities, increased coordination with regulatory bodies like SENAGUA will further guarantee its long-term sustainability. A finance-based mechanism

© ERIKA NORTEMANN, courtesy of The Nature Conservancy

for the conservation of primary water sources has encouraged water users to recognize the value of natural infrastructure. It's evident from the success of FONAG's initiatives that the partners are involved because they recognize the value in protecting water sources. This case study shows the importance of a mobilizing instrument that encourages participation and designates roles to each stakeholder involved in water management.

REFERENCES

Bréthaut, Christian, and Rémi Schweizer. (2017) A Critical Approach to International Water Management Trends. Springer, 2018.

Echavarria, M. (2001) FONAG: The Water-based finance mechanism of the Condor Bioreserve in Ecuador. *In:* Mobilizing Funding for Biodiversity Conservation: A User-Friendly Training Guide - Biodiversity Prospecting. Convention on Biodiversity. https://www.cbd.int/doc/nbsap/finance/CaseStudy-Water_Quito-Ecuador_Nov2001.pdf

FONAG. 2019. Fund for Water Protection – FONAG http://www.fonag.org.ec/web/?page_id=1580 (Accessed May 3rd, 2019).

Herndon, L. (2014). Water Funds: Financing FONAG in Ecuador. http://efc.web.unc.edu/2014/07/18/water-funds-financing-fonag-ecuador/ UNC School of Government, Environmental Finance Centre (Accessed May 2nd, 2019).

IUCN and IWA (2014). Financing Natural Infrastructure, Quito, Ecuador. http://waternexussolutions.org/ContentSuite/upload/wns/all/Case_20study_FONAG.pdf

Water Funds (2017) The Quito Water Conservation Fund (FONAG). https://waterfundstoolbox.org/regions/latin-america/quito-water-fund (Accessed May 2nd, 2019)

SOCIETA METROPOLITANA ACQUE TORINO (SMAT)

"The more we make these solutions known, the more our authorities will take them into consideration, instead of promoting investments in grey infrastructure. We have to demonstrate that they are viable and that they have a lower environmental impact. The only way we can convince them is to show them. The regulation will follow."

Armando Quazzo, Development & Marketing Manager, SMAT

LOCATION
Torino, Italy

POPULATION SERVED
2.2 million

NATURE-BASED SOLUTIONS
5. Urban green infrastructure including green roofs, spaces and water harvesting

REGULATORY DRIVERS
Water quality standards

Societa Metropolitana Acque Torino (SMAT; formerly Azienda Acquedotto Municipale)) is a provider of water supply, sewer management and wastewater treatment for the Metropolitan Area of Turin, located in northwestern Italy. Since the 1960s, the utility has relied on mountain springs, wells and surface water from the River Po to deliver drinking water to 289 municipalities and over 2.2 million inhabitants (SMAT, 2019). SMAT extracts raw water from several sources, each characterized by varying quality and reliability. Raw water from mountain springs is of high quality but seasonal variations result in fluctuating availability. Wells provide consistent quality water but are vulnerable and difficult to relocate in the event of soil contamination from agricultural and industrial pollutants. Surface water, which contributes to about 25% of SMAT's drinking water supply, also suffers from seasonal variations and is vulnerable to pollution. In the early 1990s, SMAT shifted their raw water abstraction point to the River Po's upstream branches to acquire a better quality supply of water. Despite this change, the utility faced high operation and maintenance costs as their largest drinking water treatment plant was required to perform multi-stage treatments to combat contamination.

Under pressure to service the population of Metropolitan Turin while being reliant on an inconsistent and poor quality water supply, SMAT's senior leadership needed to devise a long-term, resilient water management strategy. The situation called for a more creative approach, one that went beyond simply meeting regulatory compliance around water quality. In 2002, SMAT designed a circular economy project that would use natural infrastructure as a means of filtering out contaminants. The creative approach to water quality threats is an inspiring reminder of the vital role utility leaders must play in mainstreaming NBS.

SMAT's Raw Water Lagoon was designed on the premise that increasing storage capacity through reservoir construction would ensure water availability and improve quality for the residents of Turin. The unconventional project involved refurbishing inactive gravel quarry sites into reservoirs to access a cleaner and more reliable source of water than that extracted from the River Po. SMAT engaged with their local water regulatory authority, Autorità d'Ambito Torinese, who was responsible for reviewing the initial feasibility study. The project proposal required a two-stage review and approval process from both the local and central regulatory authorities. The strength of SMAT's business case relied on the principle that the Raw Water Lagoon project would not incur more costs than the existing cost of treatment. Capital expenditures would increase with the construction of the reservoir and pumping station, but in the long-term SMAT could justify a cost reduction using natural infrastructure instead of chemical inputs. The initial review stages were greeted with a high level of interest and requests for additional information from the local government (Autorità d'Ambito Torinese). SMAT recognizes that early engagement with authorities regarding the project design was advantageous as it helped the project pass smoothly through later stages of authorization.

The project's construction involved reinforcement of gravel quarry's slopes and creation of a new pumping facility to deliver the reservoir's raw water to the existing treatment plant, located 7km away. While it may seem inefficient to transport the water, SMAT calculated that these additional costs were well absorbed when compared to the operational costs of chemicals and filter replacements required to treat dirty water from the River Po. The reservoir's water is guaranteed to be of higher quality since it undergoes a process of natural filtration while passing through the gravel quarry walls. The utility has additional time to monitor and detect contamination levels in the River before they reach the reservoir supply, rather than having to react immediately to contaminated water pumped directly from the River Po.

Seeing immediate results from the lower chemical inputs and energy usage in their treatment plant, SMAT was convinced to expand the capacity of the Raw Water Lagoon. They have plans to construct a second reservoir by 2024 (IWA/TNC, 2018). The new reservoir will be connected to the existing one, which serves to increase resiliency and avoid the cost of constructing another pumping system. In addition to promoting stakeholder engagement and awareness in this next phase of development, it will be key to strike a balance between the economic and technical prerequisites, notes Armando Quazzo, Development & Marketing Manager at SMAT. For example, it was crucial that SMAT acquired the gravel quarries at the end of their productive lifespans, to guarantee a good price and promote the sustainability of a project that would truly revitalize an idle site. Entering negotiations for these properties at the right moment was an important consideration in the financial analysis of the project.

The slow authorization process for SMAT's Raw Water Lagoon project is an important consideration when analysing the potential of NBS programs in Italy. Bureaucratic delays are a great hindrance to promising infrastructure projects and can hinder momentum for utilities that have acquired funding and prepared plans. SMAT claims it took a total of 720 days from the date the project was conceptualized to the construction start date. This procedural barrier is an area for evaluation, particularly when considering the potential of infrastructure investments or projects (either NBS, hybrid or grey infrastructure) to address urgent water threats.

Climate change's role in increasing global temperatures threatens the long-term operation and maintenance of SMAT's reservoirs. Southern Europe has experienced higher temperatures over the past few years, which can

Aerial view of SMAT's raw water lagoon © SMAT

result in a proliferation of algae in the reservoirs. This issue presents a risk to the reservoir's capacity since it would increase weed growth in the reservoir and subsequent treatment costs for the utility. However, these costs are marginal and in the worst-case scenario of extreme drought or a poisonous substance spill, SMAT would be able to rely on their original abstraction points along river.

The Raw Water Lagoon project transforms an otherwise unproductive gravel construction site into a valuable resource for obtaining high quality raw water. The project improves SMAT's operational resilience in the face of increasing water scarcity. SMAT calculates that the construction of these two reservoirs guarantees an additional storage capacity of 8 million m3 in raw water supplies (their first reservoir can withstand a volume of 2 million m3 and the second is expected to supply an additional 6 million m3 water) (IWA and TNC, 2018). "With the construction of a second reservoir, we can ensure the availability of water for the City of Turin, with 1 million inhabitants, for at least two months" explains Armando Quazzo.

In 2011, Italy's regulatory environment underwent significant changes which changed the structure of investments and tariffs by assigning responsibility for regulation and control of water services to ARERA, the Italian Regulatory Authority for Energy, Networks and Environment. With responsibility to prepare and update the tariff method for determining fees for integrated water service, ARERA can play a role in incentivizing utilities to adopt NBS. Regulatory stimuli can prompt utilities to consider NBS as an option. ARERA has indicated a desire for international collaboration among water sector regulators on the topic of stable regulatory practices and frameworks within the EU. In April 2014, it promoted the launch of the European Water Regulators network (WAREG) and has held the presidency since 2015 (ARERA, 2019). The drive for collaboration is promising but will need to expand beyond the regulator network in order to scale up NBS programs for European utilities.

SMAT champions the NBS approach due to the obvious benefits of lowered costs, reduced chemical inputs and the opportunity to reclaim gravel quarries as productive natural infrastructure. Italian regulatory bodies are open to the possibilities of NBS for improving water quality, yet there is little being done to incentivize or promote these frameworks. Unfortunately, even the utility community can be hesitant toward innovation. Recognizing the essential role of scientific research in the evolution of the water sector, SMAT inaugurated its own Research Centre in 2008, whose mission is the development of innovative projects through research and experimentation (SMAT, 2019). Key areas include defining the value of water within a circular economy, the source control of pollutants and ensuring resilience toward water security risks. This in-house research capacity enables SMAT to continue innovating with pilot projects and technologies and remain on the leading edge of new approaches to water resource management.

Armando Quazzo sees SMAT's success with NBS as an opportunity for the sector, emphasizing how important it is to showcase examples of NBS to accelerate their acceptance. "The more we make these solutions known, the more our authorities will take them into consideration, instead of promoting investments in grey infrastructure. We have to demonstrate that they are viable and that they have a lower environmental impact. The only way we can convince them is to show them. The regulation will follow". Too often, utility companies see the starting point as the regulation and try to design approaches that will only get them to the end point of compliance. From SMAT's perspective, a utility leader who is able to champion the NBS approach, not just to regulators but to other utilities and the wider public, is absolutely imperative to changing this approach. Their intention to continue scaling NBS projects is an encouraging push to utilities around the world who are preparing to make the leap.

REFERENCES

ARERA (2010). Autorità di Regolazione per Energia Retie Ambiente. ARERA, the Italian Regulatory Authority for Energy, Networks and Environment https://www.arera.it/it/inglese/about/presentazione (Accessed June 18,2018)

IWA and TNC (2018). CASE 5: Water Lagooning in River Po, Italy. Relocation and water lagooning of river Po for drinking water purposes, SWOT Analysis.

SMAT (2019). SMAT Metropolitan Water Company of Turin. SMAT Municipalities https://www.smatorino.it/comuni-smat/ (Accessed June 18, 2019)

SMAT (2019). SMAT Research Center https://www.smatorino.it/centro-ricerche-smat/ (Accessed June 19, 2019)

MILWAUKEE METROPOLITAN SEWERAGE DISTRICT (MMSD)

"We were viewed as the polluter and not the protector of the environment. I didn't like that. I didn't think that was right. We had to change our image"

Kevin Shafer, Executive Director, MMSD

LOCATION
Wisconsin, USA

POPULATION SERVED
1.1 million

NATURE-BASED SOLUTIONS
1. Reforestation and forest conservation

3. Wetland construction, restoration and conservation

5. Urban green infrastructure including green roofs, spaces and water harvesting

6. Targeted land protection, including agricultural best management practices and improved soil health and monitoring

REGULATORY DRIVERS
Clean Water Act

Great Lakes Restoration Initiative

Great Lakes Water Quality Agreement

Milwaukee Metropolitan Sewerage District (MMSD) provides water reclamation and flood management services for 28 communities in southeastern Wisconsin. Their service area is fed by six different watersheds. The utility operates two reclamation facilities located on the shore of Lake Michigan, the second-largest Great Lake of North America.

In the early 2000's, the utility faced public backlash when wet weather events led to sewage overflow, contaminating local waterways and Lake Michigan and causing beach closings. The State of Wisconsin filed a complaint against MMSD in 2005 for dumping more sewage into the lake than permitted by law (US EPA, 2019). Negative perceptions of the utility were compounded by a lack of public outreach, education and a disconnect between public expectations and regulatory compliance. The public was less interested in the causes behind sewage overflows and easily swayed by media coverage and local politicians that portrayed the utility as a villain for harming the environment and posing health risks to Milwaukee residents (Rondy, 2006). "We were viewed as the polluter and not the protector of the environment. I didn't like that. I didn't think that was right. We had to change our image" explains Kevin Shafer, Executive Director of MMSD. To date, the utility has invested over $4 billion USD to reduce sewer overflows, including the expansion of a large tunnel to store and convey wet weather flows. For the past 40 years, they have maintained a stream and lake monitoring program and a sewer separation project that identifies areas where storm flow can be rerouted. Investments in grey infrastructure were a starting point, but solving Milwaukee's water quality challenges and changing public perceptions required a multi-faceted approach with residential involvement at its core.

As far back as 2001, before MMSD's system was overwhelmed by record rainfalls, Kevin Shafer proposed an idea to his fellow utility leaders to reduce water pollution and improve regional water security: sell old pickle barrels to Milwaukee residents for installation on their downspouts. The barrels could collect rainwater to reduce stormwater runoff and consumption, thereby minimizing the risk of sewage overflows. They laughed at him for "thinking so small".

MMSD has sold 24,628 residential rain barrels, which now store 1 million gallons (3,785 m3) of water.

MMSD's robust nature-based solutions portfolio integrates natural infrastructure into the utility's everyday operations and is centred on the notion that residents play a role in improving water quality. Green infrastructure projects include rain gardens, rain barrels, porous pavements, green roofs, bioswales, trees and tree boxes, as well as rainwater harvesting. Their green summer program hires and trains 30 interns each year to support these initiatives on the ground. They also host free rain barrel workshops (offering one free rain barrel per household) to teach residents about water conservation and how rain barrels prevent polluted stormwater runoff.

MMSD additionally prioritizes the preservation and restoration of natural landscapes such as forests, floodplains and wetlands. Their Greenseams program helps prevent flooding and water pollution by preserving land that contains water absorbing soils, specifically in areas expected to experience major urban development in the next 20 years. MMSD hired a national non-profit conservation organization, The Conservation Fund, to forge partnerships and manage the voluntary purchase of these undeveloped, privately owned properties. Wetland maintenance and restoration at the sites will increase water storage capacity, preserve wildlife habitat and create recreational opportunities for residents. MMSD is confident that allowing the land to better manage water flows will reduce the risk of downstream flooding. Their well-established Greenseams program is an indication of

Milwaukee residents plant a rain garden to help manage water where it falls and reduce water pollution © MMSD

the importance in connecting land management and water security in utility operational budgets. To date, MMSD has purchased over 3,900 acres of land for this program.

Milwaukee is the largest city in the highly agricultural state of Wisconsin and located at the downstream end of the Milwaukee River Basin. MMSD's water supply and catchment area suffer from water quality degradation due to excessive levels of phosphorus, sediment and bacteria from urban and rural stormwater runoff. As a compliment to the Greenseams, the Working Soils program invests in soil health to reduce pollution and improve natural storage capacity by permanently protecting privately held agricultural land in the Milwaukee River watershed floodplain. MMSD staff work with landowners to implement agricultural conservation practices that improve soil health and mitigate future flooding. Reaching a common understanding with the agricultural sector on water management was crucial as farmers faced pressures to grow more on less land and the utility struggled to reduce downstream pollutants.

MMSD reasoned that a non-political and educational attitude combined with financial incentives, would convince farmers to reduce their fertilizer and pesticide usage on the crops that contribute to polluted runoff in the rivers. Changing the perspective and building the trust of the agriculture sector involved a long educational process, but a few factors worked in their favour. Partnering with the U.S. Department of Agriculture, the University of Wisconsin and other local agencies established a solid foundation for the Working Soils program. "We needed to show the farmers there was a problem. Why change if there's no problem? We started building the scientific evidence and then showed them the sources of the problem. Farmers need clean water just like everyone else, so naturally they wanted to help", Kevin Shafer comments. The Conservation Fund works on the frontlines of partnership building and many of the Greenseams conservation areas overlap with the Working Soils communities. This overlap illustrates the value in a multi-faceted application of NBS where community engagement efforts can positively reinforce one other to accelerate awareness, acceptance and upscaling.

MMSD is regulated by the United States Environmental Protection Agency (EPA), who indicated support for NBS in 2007 by integrating green infrastructure into federal regulatory programs (US EPA, 2017). The EPA provides specific guidelines for integrating green infrastructure approaches into consent decrees, enforcement actions that are issued by a state or federal agency when an entity has failed to comply with federal environmental laws. MMSD's NBS programs were initiated on a voluntary basis and not through a court order. However, several consent decrees with green infrastructure provisions have been implemented in the U.S. since 2003, presenting an interesting regulatory avenue to help solidify NBS as a standard in utility management.

The U.S. Clean Water Act requires renewal of operational permits on a 5-year cycle. MMSD's Wisconsin Pollution Discharge Elimination System Permit expired in 2018, requiring the utility to conduct a series of infrastructure improvements. MMSD saw marginal impacts from the high cost grey infrastructure approach proposed by the EPA and the Wisconsin Department of Natural Resources to meet permit requirements. Having successfully surpassed the requirement to produce 12 million gallons of water with green infrastructure in the previous permit (negotiated up from an original 5 million gallons), they felt confident that a target of 50 million gallons for the 2019-2024 permit could be accomplished. Proactively engaging with the EPA and Wisconsin Department of Natural Resources allowed MMSD to push forward their NBS agenda and showcase their outstanding results. Since

inception, the utility boasts a total of 39 million gallons of water produced by their green infrastructure program. MMSD's attitude sends an important message to utilities struggling to prioritize NBS projects in a weak regulatory environment. Proactive engagement with regulatory authorities to formalize requirements for NBS involves acknowledging the opportunity in regulatory involvement, rather than seeing this presence as the sign of a problem.

MMSD's funding options for NBS benefit from the fact that the utility is a regional government agency and tax authority. Each year, the utility budgets and distributes stipends for NBS projects across their municipalities, based on each area's equalized property tax value. In 2018 this budget was calculated at $2 million USD, increasing to $5 million in 2019. The municipalities are required to vet their project proposals, which range from bioremediation facilities to green roofs, with MMSD for approval. MMSD designed this approach with the understanding that if they want to drive green solutions forward under tight public funding, they would have to provide financial support. They additionally founded the Green Infrastructure Partnership Program, a competitive RFP that offers incentive funding for green infrastructure strategies designed by property owners and developers.

MMSD receives financial support from the federal government through various avenues, most notably the Clean Water Fund Program and Great Lakes Restoration Initiative (GLRI). The EPA established the Great Lakes Restoration Initiative in 2010 to finance pollution monitoring, toxic waste clean-ups, invasive species management and habitat improvements. MMSD has received USD $9 million in grants from the programme, a critical funding source for their NBS programs. GLRI represents the largest federal investment in the Great Lakes, surviving threats to budget cuts in recent years.

With the goal of accelerating the pace of NBS implementation in the U.S., MMSD adapted their definition of "infrastructure" to include nature-based solutions. They have encouraged this shift at the federal level to ensure that green infrastructure initiatives can be incorporated into existing funding mechanisms. A transition appears to be taking root as the GLRI announced in early 2019 a grant opportunity for projects

This Greenseams® property just outside of Milwaukee helps reduce the risk of flooding with restored prairie features and a permanent conservation easement. © MMSD

that expand green stormwater infrastructure in Great Lakes communities. Securing confidence in NBS as a viable option for meeting water quality standards remains a challenge in the engineering community due to perceptions of uncertainty on performance and the timeframe for impact. Therefore, mandates from the U.S. EPA to support green infrastructure initiatives can have significant impact in accelerating the growth of NBS programs across the country. In urging early stage conversations with regulators, MMSD has positioned themselves at the forefront of the movement.

Kevin Shafer recognized at an early stage that changing perceptions about the utility at a local level would involve more than statements. MMSD needed to demonstrate action with its promised commitment. Drawing the connection between the quality and quantity of Milwaukee's water and land stewardship has called into play a broader set of actors to consider their impact on water sources. The utility has strategically invested in water management through a process of outreach, engagement and education, tying NBS projects to the local community and agriculture sector. MMSD's programmes subsequently demonstrate the substantial co-benefits that NBS can deliver beyond those related to hydrology.

REFERENCES

Rondy, John (2006). Shepherd Express. "MMSD works. So why does everyone hate it?". August 24-30th, 2006. www.shepherd-express.com

US EPA (2019). Report to Congress on Combined Sewer Overflows to the Lake Michigan Basin. Available at: https://www3.epa.gov/npdes/pubs/cso_reporttocongress_lakemichigan.pdf

US EPA (2017). Green Infrastructure. *Integrating Green Infrastructure into Federal Regulatory Programs.* https://www.epa.gov/green-infrastructure/integrating-green-infrastructure-federal-regulatory-programs (Accessed May 29th, 2019)

GHANA WATER COMPANY LIMITED (GWCL)

"Regulatory frameworks to support catchment management are in place, but not effectively enforced for compliance, which has led to water quality degradation at most of the utility's abstraction points. Compliance needs to be ensured to protect public health, water security and reduce production costs for the water utility"

Mark Ayertey, Water Quality Officer, GWCL

LOCATION
Accra, Ghana

POPULATION SERVED
10.5 million

NATURE-BASED SOLUTIONS
1. Reforestation and forest conservation

2. Riparian buffers or restoration

6. Targeted land protection, including agricultural best management practices and improved soil health and monitoring

REGULATORY DRIVERS
Establishment of Water Resource Commission and River Basin Boards

Nature-based solutions have significant yet untapped potential in Ghana. Ghana Water Company Limited (GWCL), the main provider of urban water supply in Ghana, recognizes an important role for natural infrastructure in the attainment of Sustainable Development Goal 6: to ensure availability and sustainable management of water and sanitation for all. As regulatory frameworks for managing water resources continue to be strengthened, GWCL's strong motivation to apply nature-based solutions in catchment management is clearly manifested through their active engagement with water resource agencies and users across the country.

GWCL's raw water supplies are derived from rivers, lakes, reservoirs and groundwater, which need to be sustainably managed to meet the demands of an ever-growing population. The amount of water availability in Ghana can change drastically from season to season. As a region vulnerable to the impacts of climate change, GWCL is aware that their hydrologic systems face the increasing threats of variable rainfall and recurrent droughts. The susceptibility of their water sources is exacerbated by the damages caused from human activity, such as illegal logging and mining, deforestation, pollution from liquid and solid waste disposal and poor agricultural practices. These harmful activities degrade the vegetative cover along the banks of water sources,

Students, GWCL staff and WRC staff plant trees within the catchment of Barekese dam for World Water Day 2017 ©WRC

leaving them even more exposed to the impacts of climate variability. Poor water quality and insufficient quantity have direct implications for GWCL's operations. GWCL faces critical water quality challenges caused by pollution from effluent discharges and the poor siting of waste management areas. In the past, the utility was forced to suspend operations at their Nsawam treatment plant (outside of Accra) for several months due to the high costs of treating the turbid raw water.

Population growth, economic development and changing consumption patterns are placing strains on Ghana's water supplies. GWCL's motivation to tap into ecosystem services as a means of regulating and improving water quality is driven by the potential for high rewards. Integrated approaches to water management have proven successful over the past few years, encouraging GWCL to prioritize collaborative efforts with water agencies and improve awareness on the topic of NBS.

A series of reforms to Ghana's water sector in the 1990s decentralized responsibilities for water and sanitation provision. The reforms aimed to improve coordination and collaboration throughout the water sector by delegating responsibilities across several agencies: the Environmental Protection Agency (EPA) was tasked with ensuring water operations would not harm the environment, the Water Resource Commission (WRC) would provide regulatory and water resource management oversight, and the Public

Utilities Regulatory Commission (PURC) would establish tariffs and quality standards for the operation of public utilities. The Community Water and Sanitation Agency (CWSA) was established to manage rural water systems and GWCL was delegated responsibility for urban water supply (GWCL, 2019).

The establishment of bodies such as the WRC have improved cooperation in Ghana's water sector, an important element for the future growth of NBS. The WRC acts as the central coordinating body for water resource management at both a national and local level. At the national level, it is focused on strategic engagement with major water users like GWCL and CWSA and water-related regulatory institutions, data management institutions and NGOs. At the decentralized level, the Commission operates through an administrative framework with coordination bodies known as River Basin Boards. Thus far, WRC has supported the institution of 7 River Basin Boards, which are composed of selected stakeholders with key roles in addressing the water resource challenges of each basin. WRC contributes to the development of targeted action plan interventions for the rehabilitation and preservation of these important water bodies. This is accomplished through capacity building workshops, public meetings and awareness raising efforts.

Ghana's River Basin Boards have been effective platforms for stakeholders to identify and propose solutions to the context-specific issues in each region, ultimately working toward the development of an integrated water resource management and investment plan (WRC, 2019). Collaboration with River Basin Boards has provided an entry point for GWCL to participate in catchment protection efforts and contribute to the developing management plans. As a major water user, GWCL has a representative on the WRC board and each River Basin Board in the country. This affords the utility a level of representation in every WRC project, as well as a voice in the formulation of laws and policies. The close working relationship between utility and regulator is further evidenced by the establishment of an internal department in GWCL, specifically tasked with liaising with WRC to prevent pollution around the utility's raw water abstraction points. Another example of coordinated efforts toward catchment management is illustrated in the water fee structure. The first regulation developed and adopted by Parliament under WRC was the Water Use Regulations Legislative Instrument (L.I.) 1692 (2001). It serves to regulate water use permits or water rights for various water and allocates fees toward catchment management activities including reforestation of degraded water sources, public awareness and education, and ecological monitoring.

The WRC can incentivize the use of NBS for water management through existing laws and regulatory frameworks. For example, Section 35 of the Water Resources Commission Act 522 stipulates that regulation can be made for the protection of watersheds or for preserving existing uses of public water (WRC, 1996). In addition to using existing mandates, WRC has also pursued innovative policies. In 2004, the WRC partnered with the Ministry of Water Resources Works and Housing, along with other stakeholder institutions and interest groups, to devise a consolidated Buffer Zone Policy that would address environmental degradation in the region and outline objectives for more sustainable practices (Ministry of Water Resources Works and Housing, 2013). The Buffer Zone Policy represented an important step toward a national policy on buffer zones for river basins by instituting a set of procedures to control harmful catchment activities. The policy has not been formally enacted into legislation, which complicates efforts to enforce compliance.

While the frameworks advance sustainable approaches to catchment management, GWCL attests that lack of enforcement means water quality issues often remain unsolved. "The key issue is enforcement for compliance. Water quality challenges are prolonged as a result of limited or weak enforcement of regulations on effluent, waste and wastewater management, as well as other harmful catchment activities" explains Mark Ayertey, Water Quality Officer for GWCL. There are indications that enforcement efforts are taking root at a local level. For example, registration of water users under section 11 of the Water Use Regulations LI 1692 of 2001 is carried out by the local authorities, who additionally monitor encroachment and improper waste disposal. Nevertheless, without formal legislative backing, enforcement is dependent on strong local governance structures.

GWCL recognizes a barrier to widespread adoption of NBS in Ghana is the lack of understanding and confidence in these approaches at both a policy level and within the utility. Overcoming this barrier requires working closely with River Basin Boards to increase knowledge about how NBS can improve water management and the associated costs and benefits when compared to grey infrastructure. GWCL has creatively turned to Water Safety Plans (WSP), an approach of managing water supply from catchment to consumer, to illustrate how NBS can support the delivery of safe and secure water supplies. GWCL has developed WSPs for a number of their treatment plants and actively engages with customers through the implementation of these plans to demonstrate impact. A conscious effort to secure the participation of different community stakeholders, local authorities, and non-governmental organizations, and educational institutions will be necessary to accelerate adoption of NBS.

Increasing public awareness of water quality issues, combined with more effective enforcement of existing regulations could have a significant impact on the water quality challenges facing GWCL. To reach this point, GWCL must focus on gathering reliable water quality data to support improved management and protection as well as design targeted public engagement strategies. WRC is also considering ways to improve education and awareness around water issues. They currently engage with the community through durbars, gatherings of community elders and residents for educational, awareness raising or communication purposes, and additionally rely on stakeholder meetings or workshops and existing advocacy learning platforms. Continuing to collaborate with different community stakeholders, local authorities, NGOs and educational institutions is critical for the long-term sustainability of NBS in Ghana.

REFERENCES

GWCL (2019) History of Ghana Water Company Limited. https://www.gwcl.com.gh/gwcl_history.pdf (Accessed May 19th, 2019)

GWCL (2019) Ghana Water Company Limited. https://www.gwcl.com.gh/company_profile.html (Accessed May 19th, 2019)

WRC (2019). Water Resources Commission http://www.wrc-gh.org/ (Accessed May 17th, 2019)

Ministry of Water Resources Works and Housing (2013). Riparian Buffer zone Policy for managing fresh water resources. http://www.wrc-gh.org/documents/acts-and-regulations/ (Accessed May 20th, 2019)

WRC (1996) Water Resources Commission Act, 1996. http://www.wrc-gh.org/documents/acts-and-regulations/ (Accessed May 20th, 2019)

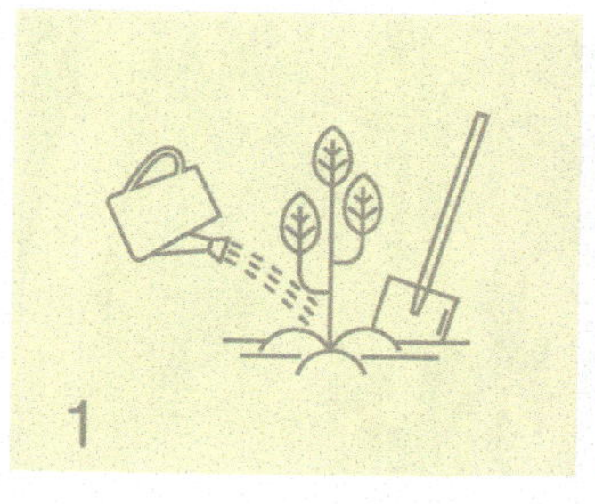

MAYNILAD WATER SERVICES & MANILA WATER

"The Annual One-Million Tree Challenge serves as a way to protect our six targeted watersheds namely La Mesa, Ipo, Angat, Umiray, Laguna Lake and Marikina River. But this can't be done over five years. We are hopeful that through this noble project and the steady support of everyone particularly our three concessionaires — Manila Water, Maynilad and Bulacan Bulk Water — we will be able to protect our vital water sources, prevent flooding and mitigate other natural disasters."

Reynaldo V. Velasco, MWSS Administrator

LOCATION
Manila, Philippines

POPULATION SERVED
15 million

NATURE-BASED SOLUTIONS
1. Reforestation and forest conservation

6. Targeted land protection, including agricultural best management practices and improved soil health and monitoring

REGULATORY DRIVERS
Concession Agreement

Clean Water Act

Philippine Water Code

Watershed management and protection in Metropolitan Manila are formally recognized as the responsibility of the Department of Environment and Natural Resources (DENR) and the Metropolitan Waterworks and Sewerage System (MWSS). In 1994, an impending water crisis stemming from water service provision shortcomings and deteriorating infrastructure forced the Philippine government to embark on a reform agenda. The Executive branch of the government urged Congress to consider the benefits of a public-private partnership to operate and expand water and wastewater services in Greater Manila. Under the National Water Crisis Act of 1995, a public bidding process awarded two concession contracts to private consortia; the first to Maynilad Water Services for the West Zone of Metropolitan Manila and the second to Manila Water Company, covering the East Zone. The transactions were regarded as the largest water concessions in the world (Dumol, 2000). This case study focuses on the experiences of Maynilad Water Services and Manila Water Company Inc. to evaluate how the concession agreement has facilitated a synchronized approach to using nature-based solutions for water security.

To better understand the relationship between MWSS and its concessionaires, it is important to consider how the concession agreement was framed. The framework is straightforward: MWSS sets the service obligations for water and waste water services in accordance with reliable customer service standards. Then the concessionaires provide the necessary investments and operational expenses to meet these obligations at an agreed and approved commensurate tariff. The MWSS Regulatory Office, a body created by the Concession Agreement, is tasked with determining if these expenditures and investments are prudently and efficiently incurred by the concessionaires in the performance obligations.

To render effective partnerships created through the Concession Agreements, MWSS recognized the importance of working closely with its concessionaries, encouraging them to adopt an integrated approach towards sustainable watershed management. With over 30 government agencies in charge of various aspects of water management, these initiatives have traditionally suffered from a lack of synchronization. For this reason, MWSS sought to implement standardized and tested approaches to watershed management and reforestation across Metro Manila.

MWSS' Water Security Legacy project laid out a comprehensive plan for watershed management and protection across the key watersheds that feed Metro Manila's raw water supply. One component of this plan is the Annual Million Tree Challenge (ATMC), an effort to reforest and regenerate watershed ecosystems through the planting and nurturing of one million trees per year in six critical watersheds. The five-year endeavor was designed in partnership with the DENR and usually surpasses its target with the involvement of Maynilad and Manila Water, who are required to report their contributions to MWSS each year.

Maynilad and Manila Water have designed their own respective programs for watershed management which are well coordinated with the DENR and MWSS's Water Security Legacy project. In the case of Manila Water,

Outplanting ©ABS-CBN FOUNDATION, INC., BANTAY KALIKASAN

approaches to watershed protection include biodiversity management, reforestation, enrichment planting, watershed monitoring, and riverbank stabilization.

The goals of this management approach are to minimize soil erosion and enhance the natural value of the area. Soil erosion is minimized through enrichment planting using endemic tree species to attain a closed-canopy broadleaf forest, which leads to improved water quality. The approach likewise promotes education and environmental awareness as forest rehabilitation activities integrate an environmental education component for employee and community volunteers by informing them about the significance of the watershed to Metro Manila's water supply.

An example of the collaboration taking place between the concessionaires and the MWSS was the establishment of the La Mesa Watershed Reservation Multi-Sectoral Management Council and its Technical Working Group. The Council is composed of individuals from MWSS, DENR, Manila Water, Maynilad, ABS-CBN Lingkod Kapamilya Foundation's Bantay Kalikasan (an NGO partner) and Quezon City Local Government Unit for the prupose of overseeing the management of the La Mesa Watershed Reservation. The La Mesa reservoir provides a storage and balance for Manila Water in supplying sufficient water for the whole East Zone Concession.

The forested areas of these watershed reservations are protected but would benefit from a stronger enforcement system. A holistic, evidence-based watershed management program should be developed to effectively address the complex problems and issues affecting these critical watersheds.

Maynilad and Manila Water recognize watershed management as an integral element of Water Safety Plans (WSP). Water Safety Plans are improved risk management tools designed to ensure safe drinking water through the use of a comprehensive assessment and management approach (see Box 2). As of 2014, WSPs are required by the Philippine Department of Health for all water service providers. As the first utility in the Philippines to launch a Water Safety Plan and have it audited by the World Health Organization, Maynilad uncovers a unique opportunity to use WSPs as communication tools to gain greater visibility around their efforts to make NBS a priority. A WSP that indicates hazards in the catchment area with potential to compromise water quality might consider using NBS as a control measure to protect a water source. So far, Maynilad has shared their WSP with about 5,000 water providers in 500 districts (WHO, 2011). Given the synchronization between concessionaires and the similar risks to water quality, Manila Water's WSP has also incorporated watershed management strategies as control measures.

While the concession model for water and sanitation service delivery provides a strong foundation for coordinated NBS efforts, the situation in Manila is not without its challenges. Maynilad's Ipo watershed protection plan was designed in 2012 and still awaits official approval from MWSS. The delay is attributed to a fluctuating political environment, management changes and the complications that arise from working across several government agencies. Given the critical nature of water quality deterioration in the region, Maynilad has pushed forward with implementing elements of the Ipo Watershed protection plan despite this delay in formal approval. Maynilad emphasizes that in order for NBS programs to be successful, they have to be prioritized by the regulator. "MWSS needs to make this a priority program. We are here as their partners. We are waiting for them" states Francisco Arellano, a Senior Consultant and former Senior VP for Corporate Quality, Environment, Safety and Health. The consequence of not being able to implement a fully integrated, long-term program is the inability to accurately measure consistent or positive results from their reforestation efforts.

In March 2019, the La Mesa Reservoir reached its lowest level in 12 years due to changing patterns of rainfall in the past years. Thousands of households across the East Zone concession of MWSS began to experience intermittent supply and lower water pressure. Following this incident, water levels in the Angat and Ipo Dam dropped and followed the trend of water levels experienced during the 1998 and 2010 El Nino episodes. The event prompted the National Water Resource Board (NWRB) to reduce its allocation for irrigation and MWSS to preserve the levels of Angat for domestic use. While the crisis prompted blame, criticism, and controversial discussions about planned infrastructure projects, it also underscored the critical need for the concessionaires and their regulator to work even harder to devise an equitable and integrated solution that recognizes the responsibilities of all stakeholders across the water supply value chain, as well as a new water source and an integrated watershed masterplan.

Each case of utility reform is specific and water privatization can be a contentious topic for utilities in developing countries. Manila's concession model may have areas for refinement, but it undoubtedly establishes a unique platform for dialogue between utility operators and regulatory bodies. Contracting water service providers in this fashion has allowed for validation and prioritization of investments in water source and watershed management as part of capital and operating expenditures. Multi-stakeholder partnerships are key driving forces for NBS programs that involve conservation or reforestation elements. In Manila, a collaborative environment has come to life where the regulator and public utility pooled resources, skills, and knowledge to address a common challenge.

Looking at the level of validation and synchronization across catchment management efforts, can the public-private partnership or concession model be credited with accelerating uptake of NBS? MWSS' promotion of watershed protection plans arguably incentivized strong performance from concessionaires. Manila Water and Maynilad recognized, even at the onset of privatization, that it would be to their advantage if they participate in programs and provide funds aimed to nurture the Ipo, La Mesa, and Marikina watersheds, alongside the Angat watershed. However, there is no explicit mention of watershed protection in the concession agreement and no formal requirements or regulatory driver that pushes Maynilad or Manila Water to dedicate funds toward catchment protection initiatives. MWSS' efforts supplied initial momentum and indicate the potential for upscaling NBS through stronger regulatory frameworks. As things currently stand, the imperative nature of water quality and quantity issues in Manila demands action and a long-term vision.

REFERENCES

Dumol, Mark (2000). The Manila water concession: an insider's look at the world's largest water privatization. Available at https://ppp.worldbank.org/public-private-partnership/sites/ppp.worldbank.org/files/documents/Key%20Government%20Official%27s%20Diary_EN.pdf

World Health Organization (2011). The Philippines. Available at: http://www.wpro.who.int/topics/water_sanitation/wsp_case_study_phl.pdf

SKANDERBORG FORSYNING

"Climate change projects are an opportunity to adapt using nature to deal with increased rainfalls and prevent the flooding of urban areas"

Jesper Brix Kjeldsen, Project Manager, Skanderborg Forsyning

LOCATION
Skanderborg, Denmark

POPULATION SERVED
61,160

NATURE-BASED SOLUTIONS
4. Flood bypasses, green infrastructure for flow regulation

REGULATORY DRIVERS
Co-financing Climate Adaptation Projects through the economic regulator for water supply companies

The Danish water utility Skanderborg Forsyning affirms that when it comes to water security in a changing climate, using nature-based solutions at the local level was never a question. "We use the best long-term solutions we can find. The fact that climate change adaptation actions can support natural habitats and biodiversity is an added bonus. For the most part, nature-based solutions or green infrastructure are the most efficient at filtering rainwater, addressing flooding issues and adapting to local circumstances", explains Project Manager Jesper Brix Kjeldsen, the expert responsible for climate change adaptation at Skanderborg Forsyning.

In recent years, the municipality of Skanderborg, located in Eastern Denmark, has experienced increasing and more frequent rainfall resulting in extensive flooding of urban areas. Rainwater accumulation can lead to sewerage overflow and surface water quality degradation, impacts that will be exacerbated by the onset of climate change. Skanderborg Forsyning has faced public criticism for the poor water quality in lakes, providing an incentive for greater community engagement and targeted efforts to unlock the full value of water. Consequently, the Municipality has developed a climate change action plan to identify vulnerable areas and address flooding.

Within the Municipality's climate change action plan, Skanderborg Forsyning is involved in 12 climate adaptation projects, many of which have NBS elements. The Municipality is considered the project owner or lead and involves the utility as project supervising engineers. In practice, this is a joint partnership, where the Municipality finances the green infrastructure or NBS through taxes and the Municipality and utility apply for co-financing from the Secretariat for Water Supply or "Forsyningssekretariatet". The Secretariat is the Danish economic regulator for water utilities under the Danish Competition and Consumer Authority (Danish Competition and Consumer Authority, 2019). If the utility's request for funding is approved, they are permitted to allocate funds from the water tariff to finance the climate change adaptation project, enabling a co-financing model supported by utility and consumer. The regulatory details are available in the *Guidance on wastewater companies' co-financing of climate projects* which is anchored in the Danish Competition and Consumer Authority (Forsyningssekretariatet, 2017). The projects developed in partnership between the utility and Municipality capitalize on the capabilities of citizens and the Municipality. The involvement of Skanderborg Forsyning demonstrates how a water utility can address climate threats by investing in their system through NBS and creating business opportunities with beneficial societal outcomes. Although these projects are not explicitly referred to as "nature-based solutions", they are inherently nature-based and capitalize on the potential of natural infrastructure like lakes to capture and store rainwater, and forests to protect the water supply.

The development of several recreational parks with rainwater lakes and forests has been a central focus of Skanderborg Forsyning's work with the Municipality. The strategic positioning of these projects enables them to have an impact on flooding threats and support climate resiliency, while simultaneously delivering benefits to the surrounding community. In the case of Låsby Lake Park, Skanderborg Forsyning transformed a former industrial site into a shared community space that could accommodate excess rainwater and protect Låsby's 2,000 citizens against increasing flood risks (Aquaglobe, 2019). This area's downward sloping terrain made it susceptible to intense flooding even during brief bursts of rainfall.

Residents had expressed a pronounced desire for an area to congregate and their ideas and active participation played a decisive role in the design of the park. At the heart of the park is a rainwater lake, which takes on different shapes depending on the level of rainfall and the depth of the lake. The lake's steep and flat shores allow different entry points for residents, even when water levels are high. There are numerous paths that provide access from different parts of the town and a bridge that connects the town's eastern and western regions. These design elements, including the carefully selected colour palettes for sculptures and athletic structures, reinforce a central theme of the park: accessibility for all. Skanderborg Forsyning worked closely with a citizen group of Lasby residents from the conception phase in January 2013 through completion in 2016. Close to 900 ideas were submitted by residents during the project proposal phase and more than 100 residents attended each association meeting. The citizen group has even signed an agreement on the park's management and maintenance to ensure long-term success. Skanderborg Forsyning is hopeful that this project will inspire others to deviate from traditional roles and be part of new collaborative efforts.

Close collaboration with the technical and environmental administration of the Municipality of Skanderborg and city council has been crucial to the success of Skanderborg Forsyning's projects. To be funded as a climate adaptation project and co-financed by utility companies, projects must fulfill several criteria and illustrate the cost efficiency of these options compared to "ordinary" solutions. Skanderborg Forsyning presented a cost comparison using gravitational flow models that evaluated the difference between constructing underground pipes and pumping systems versus developing rainwater lakes that would capture and direct the water above ground. Across the 12 different co-financed projects, they demonstrated an average savings of 72%, which included using NBS as part of their climate adaptation approaches. The Municipality of Skanderborg is involved in all projects dealing with water above ground that involve use of roads and public property, like Lasby Lake Park While Skanderborg Forsyning staff serve as supervising engineers, the Municipality provides valuable internal capacity for climate adaptation projects by employing a designated climate coordinator. Equally important to securing a

Rainwater lakes at Lasby Lake Park © SKANDERBORG FORSYNING

strong partnership with the Municipality is maintaining involvement with city councils. Skanderborg Forsyning is a private utility yet it is 100% publicly owned, which means that the Municipality acts as both an owner and partner of the project. As local residents, city council members are often experiencing the impact of flooding first hand. If they are not aligned on utility efforts along with the Municipality, then Skanderborg Forsyning must bear the financial burden. Ensuring all parties are aware of the life cycle assessment will help them see beyond the rising cost of managing water.

In addition to the projects with the Municipality, Skanderborg Forsyning is part of a 6-year EU LIFE initiative called the Coast to Coast Climate Challenge (C2C CC), which includes elements of NBS. One of the projects in the C2C CC initiative spans across municipal borders and includes development of a hydrological model of the Gudenå River, identifying possible actions to handle increased volumes of water. Several of the possible actions relate to NBS. The initiative strives to design climate adaptation projects across Denmark that turn the Gudenå River's surface water, as well as precipitation and groundwater into business opportunities (C2CCC,

2019). As a part of this, Skanderborg Forsyning has gained recognition for their work and valuable international exposure.

Between 1980 to 2017, the cost of extreme weather and climate events in Denmark totalled approximately EUR 10.5 billion. This includes the devastating floods that hit Copenhagen in 2011 from sea level rise and storms, known as the most expensive natural disaster in Europe that year (Jacobson, 2019). These events have galvanised greater action toward climate change at the national level. The government designed a financing model that prior to 2016 allowed 100% co-financing of climate adaptation projects from water tariffs if they fit the criteria established by the Secretariat for Water Supply. After 2016, utilities can only apply for 75% of costs to be covered by water tariffs. The purpose of the financing model was to ensure that the Municipality would not have to bear all the initial investments. Skanderborg Forsyning is one of several utilities in Denmark taking advantage of this co-financing model over the past few years and their progress is encouraging other utilities to recognize the importance of climate adaptation projects.

The projects being carried out by Skanderborg Forsyning are recognized as a contribution towards realising the United Nations Sustainable Development Goals and support Denmark's Vision for Water 2025. These international frameworks have provided a platform to highlight and assign value to Skanderborg Forsyning's work in the area of climate adaptation. To facilitate continued partnerships and communicate the success of these projects to other utilities and the rest of the world, Skanderborg Forsyning founded a partnership with small and large water industry companies and universities known as AquaGlobe. AquaGlobe serves as a platform to connect water sector actors with the aim of developing and testing water technologies (Aquaglobe 2019). "AquaGlobe is the culmination of our journey in learning how to work in partnership with private companies, municipalities and citizen groups, while recognizing the importance

of communication and involvement", Jesper explains. Through this platform, Skanderborg Forsyning inspires other utilities to address climate adaptation using NBS approaches and harvest the commercial, social and environmental benefits.

Skanderborg Forsyning has been able to improve community and ecosystem resilience through the design of natural spaces that reduce the impacts of increased rainfall. NBS presents a means of utilizing ecosystems and natural infrastructure to regulate floods by minimizing their impact on urban areas. Skanderborg Forsyning's approach to integrating NBS elements into their climate adaptation projects is of growing importance as the frequency and intensity of extreme weather events are projected to increase around the world.

REFERENCES

About Coastal to Coast Climate Challenge (C2CC) (2019). Coastal to Coast Climate Challenge interactive website: https://www.c2ccc.eu/english/about/ (Accessed June 17, 2019)

Aquaglobe (2019). Låsby Lake Park (Låsby Søpark) https://gis-aquaglobe.maps.arcgis.com/apps/MapJournal/index.html?appid=d4d2247ee325412ca25939564fe984b5

AquaGlobe (2019). What is Aquaglobe https://www.aquaglobe.dk/en/about/what-is-aquaglobe/ (Accessed June 17, 2019)

Danish Ministry of Energy, Utilities and Climate (2019). The Climate Initiative in Demark. Available at https://en.efkm.dk/climate-and-weather/the-climate-initiative-in-denmark/ (Accessed July 2. 2019)

Danish Competition and Consumer Authority (2019). Water Regulation. Available at: https://www.en.kfst.dk/water-regulation/. Accessed August 1st, 2019.

Konkurrence-og Forbrugerstyrelsen (Forsyningssekretariatet) (2017). Vejledning om klimatilpasningsprojekter. https://www.kfst.dk/media/46132/vejledning-om-klimatilpasningsprojekter-2017.pdf. Accessed August 1st, 2019.

Tanya Jacobsen (2019). " Extreme weather events have been an expensive acquaintance for Denmark". Flood prevention and stormwater management Planning ahead and forecasting future scenarios to mitigate the consequence of climate change. Available at: https://stateofgreen.com/en/partners/state-of-green/news/extreme-weather-events-have-been-an-expensive-acquaintance-for-denmark/ (Accessed June 17, 2019)

Conclusion

As evidenced by the 10 experiences shared in this publication, utilities around the world are recognizing the untapped potential of nature-based solutions to address current and future water security challenges. This compilation of case studies provides examples on how NBS initiatives can generate cost savings, reduce carbon footprints, increase resiliency to extreme weather events, support habitat conservation or achieve all of the above, while improving water quality and availability for water users.

Several important lessons can be drawn from this diverse set of utility experiences. A prominent theme emerged around the value of NBS pilot projects, designed in conjunction with regulatory agencies, to showcase benefits and performance. Another takeaway focuses on the importance of engaging local stakeholders early on in NBS design and implementation as a means of highlighting core co-benefits and demonstrating the need for regulatory backing. In a number of contexts where the regulatory environment is nascent or undergoing reform, utilities have taken advantage of existing compliance requirements to spark conversations with their regulators about the potential of NBS. This has unfolded differently as a product of each regulatory environment, yet illustrates a diversity of ways in which NBS can achieve progress in existing frameworks for water management. In some cases, the ways in which utilities have financed their NBS interventions required identifying additional financial resources, but more often involved redirecting existing financing. These shared lessons can serve as core principles for other utilities and water users interested in the implementation or upscaling of NBS.

In addition to shedding light on the commonalities, it's important to evaluate the areas where utilities have identified opportunities for upscaling NBS. Box 3: Identified Knowledge Gaps and Table 1 captures the needs, priorities for progress and knowledge gaps expressed by each utility interviewed for the publication. They are categorized into several topics to help define the path forward and highlight the opportunities for utilities and regulators to fully embrace their roles as protectors and partners.

TABLE 1

	Knowledge Sharing	Scientific research, monitoring and capacity building	Communication and awareness	Policies and Regulatory Environments	Economic Evaluations and Finance	Collaboration
Anglian Water				✔	✔	✔
• Identify long-term partnerships, legal arrangements and regulatory regimes that support the financing and operation of NBS. • Understand and assess opportunities to better allocate public funding for landscape management in collaboration with government agencies. • Learn about case studies with successful nature-based interventions and strong engagement with third parties, including regulators.						
De Watergroep	✔	✔	✔		✔	
• Learn about standardized methods for gathering data, modelling water flows and measuring indicators to build the case for NBS. • Identify new opportunities for NBS financing. • Determine opportunities to improve public awareness and communicate the importance of protecting water resources.						
Companhia de Saneamento Basico do Estado de Sao Paulo (SABESP)		✔	✔		✔	
• Identify effective methodologies for quantifying the benefits of NBS and build the business case for inclusion in corporate strategic plans. • Identify approaches for developing an economic evaluation of NBS programs. • Learn new ways to present and share progress with other water users, customers, society, municipalities, power companies, basin committees and regulators.						
Queensland Urban Utilities (QUU)	✔	✔				
• Identify scientific studies explaining nutrient equivalency and environmental offset programs. • Learn about innovative technologies to monitor project areas such as high resolution satellite imagery. • Find tools that show the most cost-effective areas to undertake streambank restoration projects in a catchment						

More details on the Identified Utility Knowledge Gaps are found on Box 3, page 50

Utility	Objectives						
Empresa Pública Metropolitana de Agua Potable y Saneamiento (EPMAPS)	• Learn about opportunities to raise awareness among utilities on the topic of NBS. • Understand how to improve price setting to account for the cost of upstream catchment management. • Identify opportunities to solidify collaborative partnerships between local governments, community organizations, individuals residing in source watersheds and utilities.	🍃		🍃		🍃	🍃
Societa Metropolitana de Acque Torino (SMAT)	• Identify other case studies where water companies are implementing nature-based solutions. • Learn about successful approaches to raise awareness on the topic of NBS with consumers, utilities, regulators and users.	🍃		🍃			
Milwaukee Metropolitan Sewerage District (MMSD)	• Understand approaches for maintaining green infrastructure in a cost-effective manner, especially for projects located on private property. • Identify ways to incorporate NBS into the clean water funding model to ensure more sustainable, long term funding sources. • Learn how to gain greater acceptance of green infrastructure in the engineering community.		🍃	🍃	🍃	🍃	
Ghana Water (GWCL)	• Identify successful awareness raising tactics that foster greater public acceptance of NBS. • Learn how to gather rigorous scientific data and build the evidence base to convince decision-makers of the viability of NBS, specifically related to the timeframe for outcomes. • Find case studies that incorporate traditional or local knowledge of ecosystem services into assessments and decision-making.		🍃	🍃			🍃
Maynilad and Manila Water	• Identify NBS case studies related to climate change adaptation and transboundary or multi-party (political) cooperation. • Understand how to build internal technical capacity (in forestry, biodiversity and knowledge of indigenous species) to support watershed management. • Learn more about NBS approaches such as Forest Landscape Restoration, Ecosystem based Adaptation, Ecological Restoration and Protected Areas?	🍃	🍃				
Skanderborg Forsyning	• Identify the best approach for developing an economic evaluation of NBS programs.					🍃	

One of four ponds in Anglian's treatment wetland planted with native species such as purple loosestrife © ANGLIAN WATER

Identified Utility Knowledge Gaps

- **Knowledge sharing:** sharing experiences and connecting with others to improve and promote established evidence

- **Scientific research, monitoring and capacity building:** developing assessments on the performance of NBS, indicators to measure their effectiveness, and the internal technical capacity to evaluate NBS potential, implement projects and measure performance

- **Communication and awareness:** communicating the potential role and benefits of NBS to all water users

- **Policies & regulatory environments:** assessing existing or potential regulatory frameworks to support NBS

- **Economic evaluations and finance:** designing and accessing guidelines for cost-benefit analyses, cost-effectiveness and multi-criteria analysis (co-benefits) and attracting or effectively redirecting financial resources to support NBS initiatives

- **Collaboration:** building partnerships with various stakeholders (regulators, communities, private sector, NGOs, etc.)

CO-DESIGN FROM INCEPTION

The experience of several utilities suggests that emblematic pilot projects have the potential to engage regulatory agencies at an early stage and showcase the benefits of NBS in an effective way. Key factors for success in this initial stage include partnerships with local organizations and strong cost-benefit analyses. The pilot projects conducted by Queensland Urban Utilities (QUU), Anglian Water, Societa Metropolitana de Acqua Torino (SMAT) and Milwaukee Metropolitan Sewage District (MMSD) all illustrated cost savings and in some cases, have seen expectations surpass projected impact on water quality. Direct conversations at an early stage of project development with regulatory authorities and the demonstration of clear financial returns can form a regulatory basis for NBS where it was originally absent. QUU and MMSD leveraged the successful outcomes of their NBS projects to formally negotiate new parameters for green infrastructure in their operational requirements. Continued efforts to demonstrate and outline benefits of investments from design to implementation will continue to be important as Anglian Water and SMAT pursue plans to expand or replicate these projects in the coming years. In Brazil, pilot efforts like SABESP's program are helping demonstrate the potential of NBS to state level authorities.

COMMUNITY ENGAGEMENT

Many of the cases noted the value of engagement with catchment communities and customers to encourage support for NBS. Education and awareness campaigns can encourage public mandates for NBS that in turn, promote its incorporation into policies and regulatory frameworks on water quality management. It requires cultivating dialogue with local residents, communities, and landowners and stewards. For MMSD, solving water quality issues in Milwaukee was intricately linked to education efforts around stormwater runoff and the potential impact residents could have in contributing to solutions. MMSD pursued approaches that were anchored in community alliances and connected directly with local residents who invested in actions on their private property. Furthermore, community-driven NBS projects can foster a more diverse and locally adapted set of solutions to water resource management (WWAP 2018) and ensure their longevity, a finding that was clearly illustrated in the resident-driven rainwater lake projects completed by Skanderborg Forsyning. In Ghana, the Water Resource Commission is a platform to bring community engagement into the planning and implementation of NBS approaches that continues to be strengthened.

For utilities community engagement can be an effective means of demonstrating that the net beneficial impacts of their NBS programs can go beyond financial and regulatory requirements (WSAA, 2016: 19). Simultaneously, better identifying and communicating the win-win outcomes of NBS can encourage broader stakeholder engagement and improve coordination in the contexts where it is lacking (WWAP 2018: 101). As utilities continue to build their capacity in NBS, they must underscore NBS's diverse co-benefits as part of the business case for NBS.

NBS IN DIVERSE REGULATORY ENVIRONMENTS

The utility experiences in this publication reveal creative avenues for engaging with regulatory agencies and implementing NBS initiatives through existing local, national, regional or global regulatory frameworks. Communicating the ways in which utilities are promoting nature-based solutions in different regulatory contexts is an important step in facilitating uptake. For De Watergroup, capitalizing on regional legislation and policies like the European Water Framework Directive as a means to promote new governance frameworks and surface water protection policies fostered a sense of trust and built a stronger working relationship with Flemish water authorities. EPMAPS and Queensland's experiences required more interaction with sub national agencies and resulted in municipal ordinances or state level policies that help support the utility's implementation of NBS. In the case of Maynilad and Manila Water, the contractual framework of the concession agreement created an alliance between utility and regulator on catchment protection efforts. The arrangement incentivized strong performance from the utilities, while ensuring the regulator remained aware and informed about their efforts to improve regional water security.

The experiences of Maynilad and Manila Water as well as Ghana Water highlight a risk management approach often used by utilities, which can also enable communication of NBS initiatives: Water Safety Plans (see Box 2). Generally, a Water Safety Plan (WSP) requires utilities to consider and address hazards to water supplies and then design measures to mitigate these hazards. NBS can be viewed and included as long-term solutions for addressing water threats. For example, the protection of catchment vegetation can improve natural filtration and water quality, and potentially offset the need to construct additional water treatment plants or rely on costly chemical inputs. Through the framework of a WSP, utilities are encouraged to consult different stakeholders along the water value chain. This expands their focus beyond the operations of their treatment plant to the wider catchment and encourages consideration of NBS as a low impact, cost efficient approach to improving or sustaining water quality and availability. A WSP can incorporate programmes that support NBS and offer an avenue for utilities to communicate the impact and contribution of NBS to water security.

FINANCING FOR NBS

Utilities across the globe are challenged with limited budgets, yet pressures of climate change and population growth add further importance for ensuring budgets are channelled to priority investments. This requires developing business cases which recognize the value of water resources, as well as ensuring water operators can attract financing towards these initiatives. Queensland Urban Utilities' experience suggests a strong business case can move a conventional capital works project towards an NBS approach around restoration in the catchment.

In the 2018 WWDR, current direct investments in NBS approaches are less than 1% globally (WWDR 2018). Financing sources and mechanisms which support utilities in implementing NBS programs can help ensure NBS complements conventional, built infrastructure investments in the water sector (Browder et al., 2019). The experience of EPMAPS illustrates how governance frameworks like payments for ecosystem services (PES) can help capture the full value of water resources by fostering collective action. Shifting the financial burden across stakeholders can provide a steady stream of revenue for catchment or watershed protection programs. As a government agency and tax authority, MMSD could access funding streams that enabled collaboration with surrounding municipalities and benefit from federal financial support to grow their NBS programs. Skanderborg Forsyning was able to apply in partnership with the Municipality for co-financing of climate adaptation projects that include elements of NBS, such as rainwater lakes to capture and store excess rainfall. These case studies illustrate an impressive degree of intersectoral collaboration at scale to access financing for NBS projects.

SCIENTIFIC RESEARCH, MONITORING AND CAPACITY BUILDING

Designing and managing an NBS intervention in a scientifically rigorous manner can present a considerable obstacle for utilities lacking internal capacity and expertise on the topic. For QUU, incorporating a scientific foundation into their business case was a crucial factor in convincing Queensland regulators that restoring streambanks would be more cost effective for addressing sediment and nutrient pollution than sewerage treatment plant upgrades. MMSD also found this approach successful in their engagement with the agriculture sector, where presenting the science behind pollution and runoff issues was an important building block for partnerships. Preparing and presenting the science on how NBS can address water variability requires sufficient internal technical expertise and capacity, something that not all utilities can deliver within their own capacity or in house. However, several cases demonstrated opportunities in acquiring expertise from outside of the utility to support NBS initiatives. In the case of Skanderborg Forsyning, the Municipality of Skanderborg provides valuable internal capacity for climate adaptation projects by employing a designated climate coordinator. SMAT's established water technology research centres provides in-house research capacity to be applied in pilot projects and innovations around water management. Skanderborg Forsyning has an applied approach and has set up a platform to connect actors across the water sector to develop and test water technologies. In the case of EPMAPS, FONAG was able to utilize a hydrologic monitoring program to communicate and improve outcomes of watershed investments by working closely with academic institutions. While these utilities have identified several pathways to access needed expertise for their specific initiatives, building internal capacity and accessing the tools and resources to prepare a scientifically-informed basis for NBS proposals remains a challenge for water utilities.

COMMUNICATION AND AWARENESS

Another area for further development expressed by the utilities in this publication is related to the design of effective communication and awareness strategies to illustrate the benefits of NBS. Several cases noted that an effective awareness raising campaign can help with gaining greater acceptance among regulators, water users and fellow utilities - a needed foundation for pilot efforts and long-term sustainability of these investments. Anglian Water's wetland treatment site, support from residents, environmental experts and a charitable NGO prompted the design of a pilot project that would ultimately give Anglian Water the confidence to propose dozens of additional NBS interventions in their next 5-year business plan. Ghana Water shared the important connection between awareness raising campaigns among local residents and increased recognition and abidance of existing water quality regulations. One of the main barriers to widespread adoption of NBS in Ghana is the lack of understanding and confidence that these options can deliver the same results as grey infrastructure. This is related to challenges with improving the scientific evidence base and also linked to how NBS interventions are communicated at both a policy level and within the utility. De Watergroep also noted opportunities for improved communication with farming communities and is designing an informal cooperative program to build the foundation for dialogue. Facilitating consultation processes and implementing best practices for inclusive, participatory dialogues are important areas for utilities to share tools and lessons.

REGULATORY ENVIRONMENTS

This publication showcases creative efforts on the part of utilities and their partners to maximize nature's potential to secure water supplies and improve water quality. However, in order to mainstream NBS, regulation must evolve to encourage and facilitate utility innovation, a requirement expressed by the leaders of SMAT, MMSD and Maynilad Water. In an era of climatic uncertainty and growing scarcity of resources, regulation can encourage utilities to evaluate new efficiencies and solutions to serve their customers in new ways. For these options to be reasonably considered, regulations must provide utilities with meaningful incentives. This requires entities with regulatory functions on water and environment to better understand the operations and concerns of the utility, recognizing the larger context beyond compliance with certain water quality and flow parameters. To understand how utilities might be incentivised to adopt NBS, it is necessary to acknowledge the key threats facing watersheds from the utility perspective. Positing these solutions in the context of utility concerns can be an effective means of building support. Finally, regulators should recognize the urgency around the need to respond to new pressures and design flexible regulatory frameworks to accommodate the rapidly transforming challenges that utilities are facing (Binz and Mullen, 2007).

This transition in approach and attitude requires addressing some of the barriers in regulatory environments utilities operate in. Many of the policies and regulatory environments in the case studies explored suffer from a lack of harmonization, which often stems from historically fragmented water governance frameworks, and institutions with competing and/ or overlapping mandates. Consequently, utilities are challenged to work across different agencies and adapt to leadership changes or staff departures. Contractual and approval processes are still designed with grey infrastructure engineering guidelines in mind. Project timelines can suffer from lengthy bureaucratic approval procedures. To accommodate NBS there is a need for regulatory bodies to update processes with latest evidence on emerging solutions and update procedures to accommodate built (grey) and NBS (green) under water management policies and regulations. This would allow utilities to react and invest in solutions in NBS addressing short and long term timelines as appropriate to the catchment or watershed.

ECONOMIC EVALUATIONS & FINANCING

The quantification of NBS's benefits and co-benefits presents one of the biggest challenges for utilities looking to prioritize NBS approaches to water management. This is an important gap to address, as providing information on the added value of NBS in comparison to other alternatives is vital for securing partners, investment and policy shifts. In the case studies shared, it's clear that some utilities struggle to mobilize internal support or funding to initiate NBS projects without a formal method for quantifying returns.

The challenge for many utilities with a strong and demonstrated interest in implementing NBS is a lack of access to the needed decision support tools to evaluate costs and benefits. There are a growing number of resources available to assist utilities in the decision-making process, however some require locally appropriate data to make meaningful calculations, which can present difficulties for utilities in data-scarce countries (UNEP, 2014). The development of common criteria and standards for assessing NBS should be prioritized to transition motivated utilities to the next stage of implementation.

Several experiences do illustrate how NBS can be a cost-effective option for water utilities. This should not be overstated, as NBS at scale often require large investments and context specific factors should be considered in each assessment, including an evaluation of co-benefits (WWDR 2018: 103). Utilities such as MMSD, SMAT, Anglian Water and QUU quantified clear financial returns to using natural infrastructure to comply with water quality standards over capital expansions. That said, MMSD and Anglian Water are interested in further exploring how existing models and funds can be improved or reallocated to more effectively support NBS. Utilities that have the option to fund NBS, such as source water protection efforts, from their general operating budgets can serve to "institutionalize the concept as a core water management strategy, and weaken the idea that natural infrastructure is an unconventional approach" (Gartner et al, 2017: 52).

The utilities in this publication consistently voiced an interest in identifying opportunities to improve partnership building and intersectoral collaboration in support of NBS. For example, the role of intersectoral collaboration in accelerating NBS projects was clearly illustrated in the case of EPMAPS, where a multi-stakeholder framework for source water protection supported NBS efforts despite a nascent regulatory environment. The development of Water Funds can serve as an effective mechanism to bring together a country government, water resource authorities, private sector actors, local utilities and water users within a catchment.

The issue of maintaining green infrastructure projects located on private land, establishing partnerships with landowners to promote good land management practices, and generally ensuring catchment areas are protected from harmful activities is a common challenge among many utilities in this publication. It's clear that tensions exist between water utilities and the agriculture sector, despite their common reliance on water resources. A key ingredient for success among the utilities that have seen positive advances in their relations with the farming communities has been a simple reframing of the issue. For example, improved soil practices and protection of forested areas can prevent erosion, a threat to water quality and soil productivity. If utilities only emphasize their concerns with pollution, they miss a valuable opportunity to communicate how erosion can dangerously reduce the resiliency of farming communities (Abell, R., et al. 2017). Meaningful engagement with farmers, whether that entails targeted hiring efforts to find staff with an agricultural background or assigning specific delegates to each community, can significantly improve relationship building processes.

SHARING KNOWLEDGE TO SCALE UP NBS

Contemporary water challenges require utility leaders to be forward thinking and willing to take action, even in the absence of clear decision points or financial benefits such as those associated with regulatory driven programs (Bennet et al., 2014: 61). The absence of clear regulatory frameworks to support and accelerate NBS can serve as an impediment to progress, but these 10 case studies indicate they are not a defining factor. Strong leadership on the part of utilities can substantiate for lagging or absent regulation or even push regulators to adopt new parameters for meeting compliance. These cases demonstrate a growing awareness among utilities leaders about the value of investing in nature and reflect a clear demand for regulatory environments that both protect and utilize its services. However, reaching a point where regulation and policy can support practical NBS implementation would benefit immensely from increased collaboration between utilities and their regulators.

Knowledge sharing between water utilities and among the different stakeholders involved in water management is a crucial element to support the case for upscaling NBS. It is important that these experiences are communicated in a way that is accessible to all stakeholders, including policy-makers, regulators, engineers, utility managers and contractors, that hold responsibility for translating this guidance into practice (WWDR 2019: 103). As NBS practitioners, utilities have a responsibility to improve the knowledge base by exchanging experiences (WWAP 2018). This means sharing strategies for building a successful pilot project or crafting an effective public awareness raising campaign.

There was a clear and strong demand among utilities for more opportunities and platforms to exchange experiences, regardless of the maturity of their NBS projects. Many utilities expressed a desire to access information about other utilities implementing NBS and specifically, the nature of engagement between these utilities and other watershed actors. Publicizing examples and supporting evidence that make the business case for NBS can have a range of beneficial outcomes, from promoting private sector investment, to encouraging the circulation of NBS assessment tools and standards, to enhancing a utility's internal technical capacity. This publication aims to identify the knowledge sharing gaps to encourage and enable the sharing of experiences and serve as a reference for utilities interested in integrating NBS into their operations. In sharing examples of inspiring NBS initiatives and budding utility-regulator partnerships, this compendium of experiences endeavours to catalyse new ideas, alliances and promote action.

REFERENCES

Abell, R., et al. (2017). Beyond the Source: The Environmental, Economic and Community Benefits of Source Water Protection. The Nature Conservancy, Arlington, VA, USA.

Bennett, Drew E., Gosnell, H., Lurie, S., and Duncan, S. "Utility engagement with payments for watershed services in the United States." Ecosystem Services 8 (2014): 56-64.

Binz, R and Mullen, D. 2007. Risk-Aware Planning and a New Model for the Utility-Regulator Relationship. http://www.rbinz.com/Binz%20Marritz%20Paper%20071812.pdf

Browder, Greg, Ozment, Suzanne, Rehberger Bescos, Irene, Gartner, Todd, Lange, Glenn-Marie. (2019). Integrating Green and Gray: Creating Next Generation Infrastructure. Washington, DC: World Bank and World Resources Institute. https://openknowledge.worldbank.org/handle/10986/31430 License: CC BY 4.0

Gartner, Todd, Di Francesco, K., Ozment, S., Huber Stearns, H., Lichten, N., and Tognetti, S. "Protecting drinking water at the source: lessons from US watershed investment programs." Journal American Water Works Association 109.4 (2017): 30-41.

WWAP (United Nations World Water Assessment Programme). 2018. The United Nations World Water Development Report 2014: Nature-Based Solutions for Water. Paris, UNESCO. Available at http://www.undp.org/content/undp/en/home/librarypage/environment-energy/water_governance/nature-based-solutions-for-water.html

United Nations Environment Programme. (2014) "Green Infrastructure Guide for Water Management: Ecosystem-based management approaches for water-related infrastructure projects"

Aerial view of river and landscape. © MARIO ALVAREZ

Acknowledgement

TNC and IWA would like to thank the following people for their contributions to the various cases in this publication: Armando Quazzo (SMAT), Bill Graffin (MMSD), Bjarne Langdahl Riis (Skanderborg Forsyning), Ben Yaw Ampomah (WRC), Bert De Bièvre (FONAG), Chris Gerrard (Anglian Water), Cameron Jackson (QUU), Elmer Largo (Manila Water), Florante Panganiban (Manila Water), Francisco Arellano (Maynilad Water), Galo Medina (TNC), Jesper Brix Kjelgaard (Skanderborg Forsyning), John Emmanuel Martinez (Maynilad Water), Katrine Ringgaard Jørgensen (Skanderborg Forsyning), Kevin Shafer (MMSD), Mae Berdin (Manila Water), Mara Ramos (SABESP), Marco Antonio Cevallos Varea (EPMAPS), Margaret Macauley (GWCL), Mark Ayertey (GWCL), Marvin Lascano (Manila Water), Paul Belz (QUU), Rune Kier Nielsen (Skanderborg Forsyning), Simon Six (De Watergroep), Trine Balskilde Stoltenborg (Skanderborg Forsyning), Veronica Sanchez (EPMAPS), Yang Villa (Metro Pacific Water)

We would like to thank the following people for their inputs and review: Bianca Shead (TNC), Carolina Latorre (IWA), Daniel Shemie (TNC), Kari Vigerstol (TNC), Monica Chan (TNC) and Sophie Tremolet (TNC)

We would like to thank Vivian Langmaack and Danielle DeGarmo for taking on the challenge of shaping the paper as it is visually.

And to all who have not explicitly been mentioned but have contributed to the paper in some way, we thank you.

IWA Publishing's authorised EU representative for General Product Safety Regulations is Diane D'Arras, 15 rue Duret, 75116 Paris, France, e-mail: safety@iwap.co.uk.

Printed and bound by CPI Group (UK) Ltd, Croydon, CR0 4YY

06/07/2026

02157564-0001